Food Education and Gastronomic Tradition in Japan and France

Drawing on ethical and sociological theories of food, this book presents a new approach to food education that moves beyond nutrition-centred education.

Food education has gained increasing scientific and political importance in many countries as a promising way to change contemporary eating. However, many practices fail to address two epistemological obstacles regarding its very components – 'food' and 'education'. Food has largely been thought of from a nutritionistic viewpoint alone and the ethical issues over children's freedom of choice and well-being have been absent. This book resolves these problems by applying ethical and sociological theories of food and analysing food education in two pioneering countries: Japan and France. The book focuses on taste education and gastronomy as two key concepts which have great potential to positively impact food education. Taste education is a promising alternative to nutrition-centred pedagogy which foregrounds the experience and pleasure of eating food, creating an environment for taste sensibility and food curiosity. From taste education, the picture can be broadened to examine the role and impact of gastronomy in food education. Examining the cultural traditions of France and Japan reveals how gastronomy can impact eating habits and food cultures and how these criteria should be an intrinsic part of food education. The book concludes by constructing an integrative theory for food education that moves beyond nutrition-centred education for the benefit of one's well-being.

This book will greatly interest students, scholars, policymakers and educators working on food education, food-related issues at the intersection between nutritional and social sciences, and 'gastronomes' searching for a pedagogical guide for developing their capabilities to eat in a more humanistic way.

Haruka Ueda is a Postdoctoral Fellow at the Japan Society for the Promotion of Science and Nagoya University, Japan. He won the 2022 Young Researcher's Award of the Japanese Society of Agricultural Economics.

Routledge Food Studies

Food Education and Gastronomic Tradition in Japan and France
Ethical and Sociological Theories
Haruka Ueda

For more information about this series, please visit: www.routledge.com/Routledge-Food-Studies/book-series/RFOODS

Food Education and Gastronomic Tradition in Japan and France

Ethical and Sociological Theories

Haruka Ueda

Routledge
Taylor & Francis Group
LONDON AND NEW YORK

earthscan
from Routledge

First published 2023
by Routledge
4 Park Square, Milton Park, Abingdon, Oxon OX14 4RN

and by Routledge
605 Third Avenue, New York, NY 10158

Routledge is an imprint of the Taylor & Francis Group, an informa business

British Library Cataloguing-in-Publication Data
A catalogue record for this book is available from the British Library

Library of Congress Cataloging-in-Publication Data
Names: Ueda, Haruka, author.
Title: Food education and gastronomic tradition in Japan and France : ethical and sociological theories / Haruka Ueda.
Other titles: Shokuiku no riron to kyōjuhō. English
Description: New York, NY : Routledge, 2023. | A translation and expansion of the original Japanese edition, The Theory and Pedagogy of Food Education. | Includes bibliographical references and index.
Identifiers: LCCN 2022025197 (print) | LCCN 2022025198 (ebook) | ISBN 9781032377858 (hardback) | ISBN 9781032377865 (paperback) | ISBN 9781003341895 (ebook)
Subjects: LCSH: Food habits—France. | Food habits—Japan. | Taste—Study and teaching—France. | Taste—Study and teaching—Japan. | Food—Sensory evaluation.
Classification: LCC GT2850 .U3313 2023 (print) | LCC GT2850 (ebook) | DDC 394.1/20944—dc23/eng/20220623
LC record available at https://lccn.loc.gov/2022025197
LC ebook record available at https://lccn.loc.gov/2022025198

ISBN: 978-1-032-37785-8 (hbk)
ISBN: 978-1-032-37786-5 (pbk)
ISBN: 978-1-003-34189-5 (ebk)

DOI: 10.4324/9781003341895

Typeset in Goudy
by codeMantra

Contents

Preface

The objective of this book is to construct a theory and methods of food education that can cultivate 'good eaters (gastronomes)', built upon the in-depth comparative analysis of food education in Japan and France. The earlier version of this book, published in Japanese, *The Theory and Pedagogy of Food Education*, has undergone a great deal of rewriting, with the addition of some new chapters, so that the present version can have more universal implications for worldwide food education. I thank Showado, the Japanese publisher, for allowing me to publish this updated English version.

Two factors inspired me to go ahead with this publication. Despite an increasing worldwide interest in food education politics, a comprehensive knowledge of food education has still been absent. Moreover, it has often been confused with traditional nutrition education, particularly in the English-speaking countries. I, thus, felt compelled to present holistic insights from food education in Japan and France, two pioneering countries that have continued to serve as political and theoretical 'prototypes' for neighbouring countries, both in East Asia (South Korea, Taiwan and, partly, China) and the West (Italy, the UK, Denmark, etc.). The other reason, in addition to this geographical knowledge boundary, related to the fact that, in the earlier version, an integrative theory of food education was proposed based on emerging ethical and sociological theories in an embryotic form. However, subsequent research (on well-eating and food poverty in contemporary societies) demonstrated its empirical validity, which became a strong support for this publication.

In publishing this book, I am deeply grateful to Emeritus Prof. Yoko Niiyama, my former supervisor at Kyoto University, Japan. As a leading agricultural economist, she has advanced food ethics, consumer behaviour and risk communication research in the face of two of the most imminent issues at the beginning of the 21st century in Japan – BSE (bovine spongiform encephalopathy) and radioactive contamination after the Fukushima Daiichi nuclear power accident. The idea of applying ethical theories, particularly Amartya Sen's capability approach, to food education was originally derived from such academic and social contexts. I am also extremely grateful to Prof. Jean-Pierre Poulain for hosting me as a visiting researcher at the University of Toulouse Jean-Jaurès, France (2017–2019) and encouraging me with this publication. As a pioneer in this field, he provided me with much precious advice on the sociology of food and the cultural context of taste education and gastronomy in France. Three chapters (Chapter 7 and, partly, Chapters 6 and 9) originated more or less from joint research with these two mentors, although I am responsible for all the revisions and updates in these chapters.

This publication was supported by the Japan Society for the Promotion of Science and the Lotte Foundation within the research framework of 'Well-Eating and Food Education using Sen's Capability Approach' (21J01732), 'Food Poverty in Contemporary Society' (22K14956) and 'Philosophical Inquiry into Food Education and Textbook Development' (LF000805). I extend my gratitude to the Japan Society for the Promotion of Science, the Kōnosuke Matsushita Memorial Foundation, the Kyoto University and Education Foundation and the Global Survivability Studies Programme of Kyoto University, for supporting my field work in France. My gratitude is also extended to Hannah Ferguson, the commissioning editor of Routledge, for making this publication possible, as well as two anonymous reviewers for their constructive advice.

Lastly, I especially thank my wife, Yi-ting, and my family for their love and encouragement throughout my research.

Haruka Ueda
Tokyo

Abbreviations

ANEGJ	Association Nationale pour l'Education au Goût des Jeunes
BSE	Bovine Spongiform Encephalopathy
CA	Capability Approach
CCT	Central Coordination Team
CEG	Centro Educazione del Gusto
CNAC	Conseil National des Arts Culinaires
CSGA	Centre des Sciences du Goût et de l'Alimentation
DRAAF	Direction Régionale de l'Alimentation, de l'Agriculture et de la Forêt
EPODE	Ensemble Prévenons l'Obésité des Enfants
FAO	Food and Agriculture Organization of the United Nations
FCI	Food Consciousness Institute
GMF	Gastronomic Meal of the French
IDGE	Institut de Développement du Goût chez l'Enfant
IFN	Institut Français pour la Nutrition
INAO	Institut National de l'Origine et de la Qualité
INRA	Institut National de la Recherche Agronomique
INSEE	Institut National de la Statistique et de Etudes Economiques
LAAF	Loi d'Avenir pour l'Agriculture, l'Alimentation et la Forêt
LSG	La Semaine du Goût
MAAF	Ministère de l'Agriculture de l'Agroalimentaire et de la Forêt
MAFF	Ministry of Agriculture, Forestry and Fisheries
MEN	Ministère de l'Education Nationale, de la Jeunesse et de la Vie Associative
MEXT	Ministry of Education, Culture, Sports, Science and Technology

MHLW	Ministry of Health, Labour and Welfare
MSS	Ministère des Solidarités et de la Santé
PCA	Projet Artistique et Culturel
PES	Parcours Educatif de Santé
PNA	Programme National pour l'Alimentation
PNNS	Programme National Nutrition Santé
PRECEDE-PROCEED	Predisposing, Reinforcing and Enabling Constructs in Educational Diagnosis and Evaluation. Policy, Regulatory, and Organizational Constructs in Educational and Environmental Development
SF	Slow Food
UNESCO	United Nations Educational, Scientific and Cultural Organization
WHO	World Health Organization

Introduction

1 Food Education as an Intellectual Subject?

How should we eat? How can we become good eaters? These virtue ethics-type inquiries, which penetrate through this book, can in principle be traced to the ancient Greek philosophers. Since the Enlightenment period, however, most philosophers, from Descartes to Kant, have hesitated to think seriously about 'food' due to its familiarity and corporality. On the other hand, the inevitability of 'education (or enlightenment)' for developing human virtues has been acknowledged by most philosophers, including Hume, Mill and Kant; however, hardly any seriously deepened this inquiry to concrete theoretical and pedagogical levels, with some exceptions such as Dewey. Due to this double exclusion of food and education from intellectual realms, 'food education' has remained uniquely a technical matter rather than an intellectual (or philosophical) subject, until today.

This is not a philosophy book, but I claim that studying food education is an effective entry point to inquire what constitutes 'humanity' in the context of food. The challenge is to go beyond current nutritionistic food education by expanding its theoretical foundations, from bio-psychological to social ones, including sociology, economics, ethics and philosophy. In this book, I deepen the nature of 'food' and 'education', articulate these connections and develop an integrative theory for and propose concrete methods of food education to develop 'gastronomes (good eaters/humans)'. I believe this exploration can generate universal implications, not just for food education but also for food-related issues (health, taste, pleasure, etc.) at the intersection of nutritional and social sciences.

2 Global Movement of Food Education: Two Prototypes in the East and the West

Given the recent development of food education politics in many countries, it is difficult to choose a target of analysis because such a choice requires a solid scientific justification. I have chosen to explore food education in Japan and France. Although acknowledging the constraints due to my geographic and disciplinary background, analysing these two countries has definitive advantages for other countries as well.

DOI: 10.4324/9781003341895-1

From a political perspective, Japan was the first country to establish nationwide food education policies. With the Basic Law on Food Education enacted in 2005, food education policies were proposed as a comprehensive approach to food by integrating previous food-related educational policies (notably, nutritional education, agricultural education and home economics education). About two decades have already passed with this institutional framework in place, under which all Japanese children have received food education in schools, scholars and professionals in diverse disciplines and fields have been mobilised, and over 300 research articles have been published on this topic. Some Eastern Asian countries (South Korea, Taiwan and, partly, China) have followed this political direction and Japanese food education continues to serve as a kind of 'prototype'.

On the other side of the globe, France was the first to integrate health, agricultural and cultural policies. Following a gradual process, the country finally formulated a comprehensive food education policy by the mid-2010s. Admittedly, food education is also being increasingly institutionalised in other Western countries: In the UK, following Jamie Oliver's food education initiative around the 2000s, the School Food Plan was developed in 2013 and the learning of 'cooking and nutrition' was ultimately inscribed in the national school curriculum in 2014 (Dimbleby & Vincent 2013; Page & Hart 2017). Yet, the current transformation has occurred mainly in the school meal programme and particular curricular subjects (e.g., Design and Technology), not in a comprehensive political framework. In the US, food education is currently being addressed in issue-based policies such as school meals (Levine 2008), food stamps (Landers 2007) and dietary guidelines (Jahns et al. 2018); however, a nationwide food education policy has yet to be developed (Smelkova 2015), although the federal government's reluctance has led to the development of some innovative curricula by university scholars, notably in Columbia ('Linking Food and the Environment'), Harvard ('Planet Health') and Vermont ('Democracy in the Kitchen').

Aside from its political importance, France also makes an important theoretical contribution. Scholars in the social sciences of food (most importantly, the sociology of food) undertook a thorough and critical reflection on the limits of traditional nutrition-centred education and then proposed an alternative pedagogy for teaching the 'totality' of food, called 'taste education'. It is particularly interesting to see how this novel pedagogy was invented and has been refined in relation to the philosophical and gastronomic debate on the object-subject, body-mind, nature-culture and biological-sociological dichotomies, and the numerous endeavours to overcome such dualism. In this theoretical sense, France has continued to inspire not only Western countries (Italy, the UK, Denmark, Finland, etc.), but also East Asian countries (notably, Japan).

In sum, Japan and France can be positioned as political and theoretical prototypes in the East and the West respectively, which is the rationale for the choice I made when writing this book. Admittedly, taste education and gastronomy might be unfamiliar topics to scholars in some countries, in which the current priority is to increase the number of opportunities for (any) food education practices. Nevertheless, after arriving at a certain level of institutionalisation, every country is likely to encounter a new theoretical issue concerning how the nutrition-centred tendency can be overcome; how (overwhelmingly) diverse pedagogies can be understood in an integrative manner; and how 'humanistic' eating can be ultimately conceptualised and translated into concrete pedagogies. It is these inquiries for which the focus on taste and gastronomy and the dialectic comparisons between the East and the West can contribute some universal implications regarding food education.

3 Book Structure

To achieve this goal, the book starts by examining the socio-political content (Chapter 1) and research traditions of food education (Chapter 2) by drawing on the theory of 'research programme' used in the discipline of scientific history. This analysis identifies two fundamental problems facing current food education. The first problem relates to nutritionism, in which the total (bio-psycho-social) values of 'food' are forgotten and reduced merely to the nutritional aspect. To overcome this issue, Part I diagnoses the potential of taste education as an alternative to the existing nutrition education, being promoted mainly in France, Japan and Italy, with a special focus on its pedagogical nature (Chapter 3), educational effects on children (Chapter 4) and institutionalisation (Chapter 5). These multifaceted analyses not only demonstrate the effectiveness of taste education, but also reveal the limitation that taste education cannot be truly comprehended if detached from its cultural tradition, namely gastronomy. One of the key features in this book is to expand the range of food education research to cultural analysis that most nutritionists and dietitians have previously hesitated to perform.

Part II takes this mission of further cultural exploration and aims to obtain a comprehensive understanding of French gastronomy by analysing its historical process of genesis and transmission (Chapter 6), the social representation and eating model of the contemporary French population (Chapter 7), and the philosophy of Jacques Puisais (1927–2020), a genuine gastronome-humanist who originally developed taste education (Chapter 8). This analysis highlights the sociological and philosophical nature of taste education as a unique method that, while being conscious of the negative consequences inherent in its practice, such as reproducing social inequalities (Bourdieu) and domination of taste (Foucault), can develop children as future 'gastronomes'; that is, good eaters with humanistic capabilities for paideia (appreciation of the totality of food) and philanthrōpia (conviviality with others at table). To demonstrate the universality of this French insight, Part II also includes a comparative historical analysis of Japanese gastronomy with the French one (Chapter 10).

In a different vein, the second problem concerns the 'ethical' nature of food education, which is viewed as 'self-evident' by scholars, and thus, theoretically unexamined. Drawing on evidence about taste education and gastronomy, Part III establishes the integrative theoretical foundation for food education by applying Amartya Sen's capability approach (CA), an ethical theory for evaluating people's freedom for achieving well-being (Chapter 10). In the concluding chapter, I argue this CA-based approach can serve as a solid theoretical and empirical foundation not only for food education, but also for food-related issues in the social sciences of food, particularly well-being (or eating well) and food poverty studies.

4 Contemporary Conditions Regarding Food Education

Before moving to the main chapters, I describe two major prerequisite conditions that characterise and necessitate contemporary food education. First, we now acknowledge that the 'freedom' of individuals' eating lives (the meaning of which is specified later) has resulted in various social problems. Here, I take Japan as an example, but the implication is more or less true for many countries with food education policies. In the super-ageing Japanese society, lifestyle-related diseases account for 60% of all the causes of death and amount to 30% of the total medical national expenditure (MHLW 2021a, 2021b). The current medical system is no longer sustainable; thus, promoting food education as primary prevention is inevitable to reduce the risk of food-derived lifestyle diseases. In particular, for children, although obesity is not so troublesome, the contemporary type of 'undernutrition', in which the necessary nutrient intake (caloric, vitamins, etc.) is unachieved, is increasingly problematic (MHLW 2021c). What is perhaps surprising to the non-Japanese readers is that Japanese poverty rate is the second worst (after the US) among G7 countries, with one in seven children living below poverty line, and the inequality of eating standards is increasingly exacerbated (Abe 2008; Abe et al. 2018).

However, unhealthy eating is just one side of the coin; the Japanese food system is also in peril. The self-sufficient rate has dropped below 40%, the agricultural population has decreased at a rate of over 20% every 5 years, and those aged over 65 account for 70% of the active agricultural population (MAFF 2021). Such fragility in the food system is due to structural problems in which producers are unable to receive fair profits to cover their production costs, and the equal power balance across every distributive stage is violated (Niiyama 2020). Admittedly, food choice is a personal issue but advocating solely for this type of (negative) freedom might end up limiting our (positive) freedom in the long term (for a distinction of freedoms, see Berlin 1969). Freedom of food choice is not unlimited and should, sometimes, be negotiated through public policies, among which food education came to be regarded as a promising tool.

Second, contemporary eaters now face the 'gastro-anomic' situation (or *hōshoku* in Japanese), in which traditional dietary norms were weakened (Fischler 1990), or more precisely, replaced with too many conflicting new dietary norms (Poulain 2002a). The proliferation of these dietary norms has left contemporary eaters at a loss in terms of 'what and how to eat'. The contemporary (or late modern) world, as Giddens (1999) diagnosed, is characterised by the accelerating inflow of new knowledge and the weakened legitimacy of tradition as merely 'being traditional', which then requires every individual to think reflectively about his/her own lifestyle. This point is so true for food modernity. Consequently, many diverse forms of food education have been newly proposed to rescue these eaters from this gastro-anomic situation; yet, however paradoxically, many of these exacerbate the same condition rather than developing capable eaters. This latter point is the purpose of this book.

5 Theoretical Positioning

One author called the earlier version of this book, The Theory and Pedagogy of Food Education, 'a truly interdisciplinary study of food education' (Yamada 2021), but the 'interdisciplinarity' is too ambiguous a concept and, in the worst case, it might result in being untheoretical. To avoid this risk in advance, I sketch a group of theories employed for the following analyses.

5.1 Nutritional and Psychological Theories

If you open nutrition education textbooks, in most cases you will encounter a list of nutritional or, more precisely, social psychological theories such as, to name a few: the health belief model, the theory of planned behaviour and the PRECEDE-PROCEED model (which I shall review in greater detail in Chapter 1). In many countries, this group of theories has largely informed nutritional education policies and practices, and this book does not neglect such a legacy. This is why, in some chapters, I eagerly employ this group of theories, such as for measuring educational impacts, the development of a psychological indicator of food curiosity and the analysis of health promotion systems. However, exactly because of this methodological orientation, I am to be confronted with new problems that cannot be solved effectively with this set of theories.

5.2 Ethical Theories

Ethical theories play a central role in integrating diverse findings on food education throughout this book. At first appearance, ethical theories might seem too scholastic and even intimidating to food education scholars with nutritional-science backgrounds because the discussion about ethical issues has been absent in past research (as discussed in Chapter 2). On the contrary, these theories can provide a conceptual framework that helps to navigate decision-making about the ethical dilemmas regarding a person's well-being, freedom and human development, which most practitioners are experiencing every day on the frontline.

The prevailing view that ethical theories are not useful for food education might be partly due to the developmental process in ethics research itself: Traditional ethical theories, such as utilitarianism (Bentham and Mill) and deontology (Kant), were primarily concerned with 'what the right actions are', which, in most cases, could not provide a full account of complex real problems. However, recent critical reflection on the limits of these approaches has led to the (re)emergence of virtue ethics (Anscombe, MacIntyre, Nussbaum, etc.), which inquires 'how we should live well', just as with the ancient Greek philosophers, notably Aristotle. Many virtue ethics scholars, however, still remain obsessed with metaphysical questions and are unable to inquire about particular virtues in concrete and empirical settings. The absence of virtue ethics is also true for the 'food ethics' research (Sandler 2014; Thompson 2015) that has emerged since the 1990s, despite its new and important focus on ethical issues of contemporary eating (Ueda 2022). The only exception is Amartya Sen's CA (although Sen himself did not anticipate that his ethics would be applied to contemporary food). Along with some pioneering studies (Hart 2016; Hart & Page 2020), this book becomes one of the first attempts to apply the CA to contemporary eating lives and food education.

In addition, I owe much to John Dewey's educational ethics. Although many entry points to educational philosophies (e.g., Freire, Montessori, Steiner) are possible, Dewey's thought is worth exploring for a couple of reasons. First, what distinguishes taste education from traditional nutrition education is the previously non-theorised concept of 'experience', which can be ultimately extended by Deweyan empiricist philosophy. Second, the theoretical affinity between Sen and Dewey has already been investigated (Zimmermann 2006; Lessmann 2009), indicating that the two thinkers shared similar interactionist and developmental accounts of human life. Beyond such similarity, Dewey can also provide us with concrete pedagogical implications, which would not normally be explored in Sen's CA.

5.3 *Sociological Theories*

Simultaneously, the analysis of taste education and gastronomy is largely informed by the 'sociology of eaters'. Traditionally, food was not considered a legitimate subject in sociological studies due to Durkheim's (1895) ambiguous definition, in which food (such as in rites and ceremonies) was seen as an undeniable social fact but too close to the biological dimension. Thus, food has long been excluded from mainstream sociology and occasionally visited only as a sub-theme in the sociology of consumption (e.g., Engel, Halbwachs, Le Play) and the sociology of taste (notably, Bourdieu). After experiencing such a difficult period, the sociology of eaters ultimately emerged in the 1980s by drawing on the Maussian (1968) anthropological tradition and Morin's (1973) transdisciplinary philosophy as a new research programme to investigate food as a total social fact that comprises the bio-psycho-socio-cultural dimensions of human life (Poulain 2017). This group of researchers have disclosed the totality of subjects that were previously reduced to biological dimensions, such as obesity (Poulain 2009), pleasure (Dupuy 2013), eating model (Poulain 2002a, 2002b) and eating well (Lahlou 1995; Fischler & Masson 2008). Unfortunately, mostly due to geographical boundaries, these works are not widely recognised in food education research outside France. However, I demonstrate that various sociological theories utilised in this new tradition are useful, particularly for considering the totality of food education and complementing the CA's relative weakness in systematically understanding the social structure influencing one's capability.

It is my hope that the 'interdisciplinary' approach employed in this book can be understood as a combining of existing bio-psychological theories with emerging ethical and sociological theories.

References

Abe, A. (2008) *Kodomo no Hinkon [The Child Poverty]*. Iwanami Shoten.

Abe, A., Murayama, N., Kachi, Y. & Gan, S. (2018) *Kodomo no Hinkon to Shoku-kakusa [The Child Poverty and Inequality of their Eating Standards]*. Otsuki Shoten.

Berlin, I. (1969) Two concepts of liberty. In I. Berlin (ed.) *Four Essays on Liberty*, 118–172. Oxford University Press.

Dimbleby, H. & Vincent, J. (2013) *The School Food Plan*. http://www.schoolfoodplan.com/wp-content/uploads/2013/07/School_Food_Plan_2013.pdf (accessed in April 2022).

Dupuy, A. (2013) *Plaisirs Alimentaires, Socialisation des Enfants et des Adolescents*. Presses Universitaires de Rennes & Presses Universitaires François-Rabelais.

Durkheim, E. (1895) *Les Règles de la Méthode Sociologique*. Félix Alcan. Translated by T. Miyajima (1978) *The Rules of the Sociological Method*. Iwanami Shoten.

Fischler, C. (1990) *L'Homnivore*. Odile Jacob.

Fischler, C. & Masson, E. (2008) *Manger : Français, Européens et Américains face à l'Alimentation*. Odile Jacob.

Giddens, A. (1990) *The Consequences of Modernity*. Stanford University Press.

Hart, C. (2016) The school food plan and the social context of food education in schools. *Cambridge Journal of Education*, 46(2), 211–231.

Hart, C. & Page, A. (2020) The capability approach and school food education and culture in England. 'Gingerbread men ain't gonna get me very far'. *Cambridge Journal of Education*, 40(6), 673–693.

Jahns, L., Davis-Shaw, W., Lichtenstein, A., Murphy, S., Conrad, Z. & Nielsen, F. (2018) The history and future of dietary guidelines in America. *Advances in Nutrition*, 9(2), 136–147.

Lahlou, S. (1995) *Penser Manger : Les représentations sociales de l'alimentation*. Thèse doctorale. Ecole des Hautes Etudes en Sciences Sociales.

Landers, P.S. (2007) The Food Stamp Program: History, nutrition education, and impact. *Journal of the American Dietetic Association*, 107(11), 1945–1951.

Lessmann, O. (2009) Capability and learning to choose. *Studies in Philosophy and Education*, 28(5), 449–460.

Levine, S. (2008) *School Lunch Politics: The Surprising History of America's Favorite Welfare Program*. Princeton University Press.

Mauss, M. (1968) Les Techniques du Corps. In M. Mauss (ed.) *Sociologie et Anthropologie*. Presses Universitaires de France. Translated by T. Aji & Yamaguchi, T. (1976) *Sociology and Anthropology II*, 121–156. Kubundo.

Ministry of Agriculture, Forestry and Fisheries (2021) 2020 Census of Agriculture and Forestry. MAFF.

Ministry of Health, Labour and Welfare (2021a) Vital Statistics of Japan 2020. MHLW.

——— (2021b) Estimates of National Medical Care Expenditure 2020. MHLW.

——— (2021c) National Health and Nutrition Survey 2020. MHLW.

Morin, E. (1973) *Le Paradigme Perdu: La nature humaine*. Seuil.

Niiyama, Y. (2020) *Food System no Kozo to Chosei [The Structure and Coordination of Food System]*. Showado.

Page, A. & Hart, C. (2017) *A Report on the Food Education Learning Landscape*. Jamie Oliver Food Foundation. https://www.akofoundation.org/wp-content/uploads/2017/11/2_0_fell-report-final.pdf (accessed in April 2022).

Poulain, J.P. (2002a) The contemporary diet in France: 'de-structuration' or from commensalism to 'vagabond feeding'. *Appetite*, 39(1), 43–55.

——— (2002b) *Manger Aujourd'hui, Attitudes, Norms et Pratiques*. Privat.

——— (2009) *Sociologie de l'Obésité*. Presses Universitaires de France.

——— (2017) *The Sociology of Food: Eating and the Place of Food in Society*. Bloomsbury.

Sandler, R. (2014) *Food Ethics: The Basics*. Routledge.

Smelkova, L. (2015) *Food Education in America*. Center for Science in the Public Interest & Food Day. https://kiesel.ucdavis.edu/Food_education_report_Dec2015.pdf (accessed in April 2022).

Thompson, P. (2015) *From Field to Fork: Food Ethics for Everyone*. Oxford University Press.

Ueda, H. (2022) Food ethics reexamined: Capability approach and present methodological challenges. *Journal of Ritsumeikan Gastronomic Arts and Sciences*, 7, 251–269.

Yamada, H. (2021) Book Review of 'The Theory and Pedagogy of Food Education'. *Journal of Rural Economics*, 93(3), 344–345.

Zimmermann, B. (2006) Pragmatism and the capability approach: Challenges in social theory and empirical research. *European Journal of Social Theory*, 9(4), 467–484.

1 History of Food Education

This chapter overviews the historical context of how food education was constructed as an urging social and scientific theme. I take Japan as the primary target of analysis, mainly because the Japanese case clearly illustrates how different traditions, namely nutrition education, agricultural education and home economics education, have formed a new concept of 'food education' that was eventually institutionalised by the Basic Law on Food Education in 2005. To understand the complex dynamism of these precursor pedagogies and their supporting disciplines, I apply the approach of scientific history, such as a 'research programme' (Lakatos 1978; Berthelot 2001) and 'research tradition' (Laudan 1977), both of which share the post-positivist premise with the concept of 'paradigm' (Kuhn 1962) in a more elaborate way. As an analytical concept for the evolution of scientific history, a 'research programme' consists of *hard core* (axioms and theories that define its ontology and epistemology) and *protecting belt* (hypotheses to explain anomalies). Each precursor food pedagogy, although its hard core is implicit in most cases, can be considered as such research programme (Figure 1.1).

However, it is very important to understand that, given each research programme's stability, it would be insufficient to direct criticism against it (e.g., 'social and cultural values of food are neglected' type-criticism to the nutrition education programme), for such criticism can often be addressed by reinforcing its protecting belt (e.g., added-in theories of ecological model or food environment) and cannot, thus, refute its hard core. It is, therefore, necessary to construct a new 'rival programme' with higher explanatory factors of the phenomena in question, which is otherwise called a new research tradition with higher 'efficiency' (Laudan 1977). The ultimate goal of this book is to construct this rival research programme, comprising new theoretical foundations, to explain both the scopes of previous programmes and their anomalies.

1.1 The Earliest Tradition of Food Education

1.1.1 *Etymology*

Shokuiku literally means food (*shoku*) education (*iku*) in Japanese. This term began to be used by a small group of experts in the 1980s–1990s, but its meaning was merely a

DOI: 10.4324/9781003341895-2

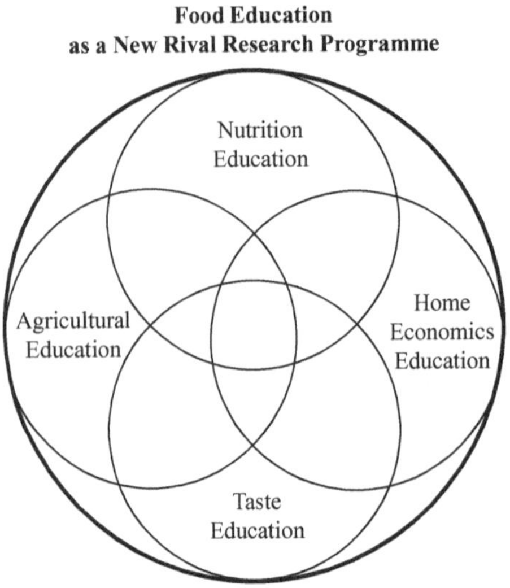

Figure 1.1 Food Pedagogies as Research Programmes
Source: Developed by the author.
Note: Taste education as the fourth programme is closely analysed in Part I, not in this chapter. Although other pedagogies (such as environmental and sustainable education) exist, these are relatively new, if compared to the three major pedagogies, and being promoted in a close relation to existing pedagogies, thus are not included in this figure.

general 'education about food and eating', with no philosophical loading. However, in the preparatory process of the Basic Law on Food Education (Shokuiku) to be issued in 2005, research into the term's etymology was conducted (such as Morita 2004) and its linguistic connection with the early thinkers of food education at the beginning of 20th century was 'accidentally' discovered. One can view, behind such a discovery, the nationalistic motivation for justifying the emerging food education policy as a 'tradition' that appeared to have its origin about 150 years previously (Mah 2010). A more positive psychology is also evident; one that attempts to extract some contemporary significance of these early thoughts as a counterreaction to the conceptual ambiguity and philosophical poverty of current food education (Kuroiwa 2007; Sato 2009; Fujii 2014). Bearing such increasing scientific attention in mind, the early thinkers of food education, notably Sagen Ishizuka and Gensai Muraki (Figure 1.2), should not be dismissed, so I now briefly review their thoughts.

Sagen Ishizuka (1851–1909) is, internationally, more well known as the founder of the macrobiotic movement than as the father of food education. Although born into a family of traditional physicians, he was obliged to study Western medicine to be an authorised doctor. During his mission as military pharmacist, Ishizuka witnessed the physical inferiority of the Japanese armies compared with Western soldiers. This disappointment later led to his publications of *Kagaku-teki Shokuyo Chojyuron* (*A Chemical Theory of Food for Health and Longevity*) in 1896 and its popular version *Shokumotsu Yojyoho* (*A Theory of Food for Health*) in

Figure 1.2 Sagen Ishizuka and Gensai Murai
Source: Portrait photographs were obtained with permissions from Food Health Ishizuka Sagen School (for Ishizuka) and Hiratsuka City Museum (for Murai). The third image was cited from Murai's (1903) *Gourmandise*, currently available in the digital archive of Aozora publisher. The image clearly represents Murai's ideal of family food education, exemplified by a knowledgeable woman teaching the other woman how to make a rational purchase and cook beef based on its anatomical illustration.

1898, the latter of which became the bestseller of his time. Ishizuka's statement in these books of 'physical, intellectual and talent education, that is, food education (*shokuiku*)' (1898, p.178) is said to be the first appearance of this term.

Gensai Murai (1864–1927) was an enlightenment thinker who wrote a series of journal columns about food, diseases and hygiene by drawing on his over two-year experience of living in the US. These columns were later assembled into the bestselling book of 1903, *Shokudoraku* (*Gourmandise*), in which he discusses that, 'the root of physical education and intellectual education is food; therefore, I am convinced that food education (*shokuiku*) is more important than these two' ([1903]2005).

1.1.2 Earliest Thoughts

The purpose here is certainly not to describe these writers' thoughts on food education in detail, so I limit myself to commenting on their ideas regarding some inherent problems in contemporary food education (notably, nutritionism, gendered ideology and nationalism), which are discussed in detail in Chapter 2.

Regarding their thoughts on nutrition, Ishizuka's approach was critical to the reductionism of modern nutritional science but pragmatic to its analytical concepts (e.g., sodium, potassium) to justify his own macrobiotic theory (Fujii 2014). In contrast, Murai was much more receptive to modern nutritional science, for his ultimate objective was to enlighten conventional family life to employ scientific thoughts on hygiene and health. Despite the primary importance attributed to nutrition, the writers' theories go far beyond orthodox nutrition education. A wide range of knowledge from Eastern thoughts on the human body and natural environments constituted the theoretical basis of Ishizuka's food education, whereas Murai's talent was reputed in his time: 'he is truly encyclopedic, not Gensai [a part of his name *ensai* has a similar pronunciation with *ency*]' (Kuroiwa 2005). As the author of *Gourmandise*, Murai's ([1903]2005) humanism is expressed in his remark that 'it is so sad that man eats but does not know how to taste' (p.105), which is somewhat akin to Brillat-Savarin's ([1825]1982) famous thesis that 'man eats; but only the man of spirit knows how to eat' (p.19).

Murai's thoughts also have implications for understanding the gendered nature of food education. In his food education, Murai envisages importing the 'frontier spirit', which he had seen in the rationality and curiosity towards new cultures among American women, into Japanese middle-class women (Yamamoto 1982). In doing so, he wanted to develop their responsibility and ability to choose foodstuffs and cook at home by themselves to be independent from their husbands (Ishida 2000). In this sense, Murai was already proposing a feminist food education that can be rarely identified in the Confucianist social atmosphere of his time.

Murai's thinking regarding family cuisine is often criticised as a nationalistic device that contributed to producing the 'nation' through family tables (Murase 2000). However, it is necessary to distinguish this 'consequence' from his primary 'intention' to develop enlightened eaters. On the other hand, Ishizuka's ideas risk being amalgamated with the exclusive nationalism of his time, for the major intention of his food education was to replace the increasingly popular Western diets with macrobiotic ones based on the consumption of rice and vegetables. In reality, his theory of food education, whether his intention or not, was mobilised in *Shokuyo Shinbun* (*Newspaper of Food for Health*: 1905–1907) to stimulate fanatic nationalism following victory in the Russo–Japanese war (Sato 2009).

In this way, these early theories help us to relativise the ideological nature of contemporary food education. In particular, universal insights for food education can be learnt from these writers' humanistic perspectives on the totality of food and rational agents. Interestingly, both Murai and Ishizuka condemned *bishoku* (gastronomy) as conducive to non-humanistic eating, a prejudice still relevant in contemporary Japan, and thus, one problem addressed in this book. Nevertheless, limitations in their educational theories should also be acknowledged. Although both thinkers aimed to educate citizens and took inspiration from Spencer's theory on 'intellectual, moral and physical education' (which must have been familiar to most intellectuals of their time) to coin the term 'food education', it is not the case that they proposed any systematic educational theory through critical reflection of Spencer's theory. Therefore, recent endeavours to extract theoretical insights into the educational nature of *shokuiku* from these writers must confront some limits. Instead of such retrospective orientation, I propose to face the complexity of current food education. As part of such a challenging but constructive attempt, I examine the definition, historical development, theoretical premise and programmatic nature of each precursor food pedagogy.

1.2 Nutrition Education

1.2.1 Definition and Historical Development

One of the most authoritative definitions of nutrition education is 'any combination of educational strategies, accompanied by environmental supports, designed to facilitate voluntary adoption of food choices and other food- and nutrition-related behaviours conducive to health and well-being' (Contento 2010, p.14). To identify the basic premises internalised in this overarching definition requires analysing its historical transition.

Modern nutritional science was officially introduced into Japan following the Meiji Restoration (1868) and became the major theoretical support for nutrition education in the replacement of traditional Chinese medicine and herbal medicine. With the prevalence of beriberi (a disease caused by thiamine deficiency due to purified rice-eating) as a turning point, the Meiji regime started institutionalising nutrition education, notably the establishment of the professional educational institution for nutritionists in 1924, the first of its kind in the world (Shimazono 1989). The post-war regime took over this project to establish numerous courses for dietitians and nutritionists in women's colleges/universities. Given its inextricable nature with other neighbouring disciplines (public health, home economics, etc.) in these educational institutions, it would be more precise to understand that nutrition education is theoretically based on a group of nutrition-related sciences in a wider sense, not the single field of basic research on nutrients and their bodily functions.

Despite its long history of institutionalisation, it was not until the 1990s that nutrition education achieved the current definition described above. Prior to this period, nutrition education, which was originally merely knowledge dissemination, had undergone three theoretical developments (Takemi 2002): First, in the 1970s, eating behaviour, which had been traditionally regarded as self-evident, became the target of serious investigation, and nutrition scientists started to explain its nature from social psychology perspectives (e.g., the health belief model). Second, as characterised by the Ottawa Charter for Health Promotion in 1986, it became a general consensus that the aim of nutrition education was to achieve not just one's health, but also a 'a complete state of physical, mental, and social well-being'. Importantly, this international framework of health promotion also helped to highlight the environmental factors regarding healthy eating behaviours. Third, targets of nutrition education had long been considered 'passive patients' who simply received dietary guidance, but in the 1990s they gradually came to be viewed as 'active learners' in response to the emergence of empowerment or participation-based educational practices.

1.2.2 Research Programmatic Nature

Drawing on these historical developments, let me summarise the research programmatic nature of nutrition education. Despite its theoretical transition, the historical consistency of nutrition education's basic premise of 'good eaters who lead healthy eating lives based on nutritional evidence' is evident. While this hard-core value has remained unchanged, new perspectives and hypotheses have been added to the protecting belt to respond to emerging criticisms and partially transform its programmatic nature: (i) from health itself to well-being as its educational goal, (ii) from passive patients to active learners as its target and (iii) from individual to socio-environmental factors on healthy eating as its analytical focus. This developmental scheme enables a better understanding of the above-mentioned definition of nutrition education today.

However, this three-layered transformation in perspectives does not mean its theoretical evolution. Indeed, nutrition education has continued to be theoretically based primarily on nutritional sciences and social psychology, the latter of which has not been updated dramatically. This point surely demonstrates the limits of nutrition education's protecting belt, but it still holds the highest explanatory power for the evolution of eating behaviours compared with other research programmes and, therefore, still serves as the most dominant programme in current food education. Given that the theories of nutrition education are highly standardised (just as the Japanese ones are quasi-American ones), this programmatic nature will be largely relevant to nutrition education in other countries.

1.3 Agricultural Education

1.3.1 Definition and Historical Development

Agricultural education, despite its cultural and regional variations, generally falls within each of the following four categories (Ueno 2014): (i) professional education for potential farmers; (ii) educational intervention with actual farmers (extension, short-term training, etc.); (iii) academic education in agricultural sciences; and (iv) social education about agriculture for children and adults. While the first three are more professional educational practices, the fourth is more general and social (Kataoka 2006), and, for this differential purpose, has also been called 'food and agricultural education' since the late 1980s. Although this food and agricultural education has a direct connection with current food education, its educational nature should be understood from a wider historical perspective, including professional education (for the hereafter historical overview, Takayama 1981).

Modern agricultural education in Japan began in the 1870s as part of the *fukoku-kyohei* (increasing wealth and military power) policy. As the theoretical basis of agricultural education, Western agricultural sciences were almost blindly introduced, which facilitated analytical approaches to farming practices to some extent, but its educational integrity as pedagogy for 'human flourishing'[1] was ultimately unachieved during the pre-war period due to numerous politico-philosophical obstacles (such as the rise of ultra-nationalism, the prioritisation of practical education for 'loyal nations' and the culturally disdained view of farmers).

Post-war agricultural education made a delightful departure under the new constitutional framework premised on the three basic principles of sovereignty of the people, respect for fundamental human rights and pacifism. To achieve the 'democratic' agricultural education (as emphasised by the American occupation), American agricultural pedagogies (home project method, school-based agricultural clubs, integrated farming, etc.) were systematically introduced. However, the next two decades (1950s and 1960s) soon saw the emergence not only of the conservative movement against this democratic educational reform, but also of the capitalist educational policy that prioritised practical education over liberal arts. Moreover, post-war economic development until the 1970s accelerated depopulation in the agricultural sector. Accordingly, agricultural education came to be increasingly viewed as an instrument for producing 'agricultural successors' by putting aside its original democratic ideal. During this period, agricultural education ceased to be professional education and was institutionalised as social education for the general public as a part of urban–rural exchange and green tourism programmes (Yamada 2016).

In addition to agricultural policy since the 1970s, educational policy has also displayed interest in the educational benefits of farming experience for children as a countermeasure to the increased incidence of delinquency, violence and bullying in school environments. A number of curricular changes (notably the establishment of '*yutori* [relaxed] education', 'life environment studies' and 'integrated studies') was implemented to facilitate practising agricultural activities (school garden programmes, farm visit, hands-on experience, etc.) to cultivate children's 'zest for living' (Yamada 2016), although this concept is void of any philosophical or theoretical justification.

1.3.2 *Research Programmatic Nature*

In contrast to nutrition education with established nutritional evidence, agricultural education has been historically vulnerable to the political climate of each period, which reduced its educational scope in the production of loyal nations, agricultural successors, students without misconduct or whatever was convenient for the state. Agricultural economics (otherwise called rural sociology) and, more recently, educational sciences have served as agricultural education's theoretical basis but failed to rescue it from these political ideologies. Nevertheless, such a socio-political nature constitutes a part of its hard core. If coupled with the original motivation of post-war agricultural education (which is still true for most scholars and practitioners in this field), the programmatic nature of agricultural education can be understood as centred on the hard-core value, whether professional or general, attached to 'good eaters who serve as democratic agents for social change in agricultural and rural environments'.

Agricultural education has recently been valorised in diverse ways across the globe (Farag et al. 2021) and Japan is not the only country that is experiencing a conflict between the values of agricultural and educational policies; the rural and the urban; and the professional and social education – and their consequent pedagogical identity crisis.

1.4 Home Economics Education

1.4.1 *Definition and Historical Development*

Prior to the birth of *shokuiku* as a political term in 2005, all educational practices related to food in schools were literally called 'Food-related Guidance' (*Shoku ni kansuru Shido*). Such Food-related Guidance has been practised in various subjects, such as home economics, life environment, social studies, science, integrated studies and more recently, school lunchtimes (MEXT 2010).[2] Among these, the first and last have substantial opportunity for practice and require more in-depth examination.

As the theoretical basis of the former class, home economics (later called human ecology) was introduced from the West into Japan as the central theme of women's education for developing a 'good wife and wise mother'. In contrast to nutritional and agricultural sciences, home economics during the pre-war period was merely an assemblage of neighbouring disciplines, and, due to the absence of its unique approach to 'family life', it ran the risk of being downgraded into sub-disciplines of relevant themes, such as food, clothing, education and housing (Kimoto 2005). Moreover, in the midst of the post-war diversification of family structures, home economics scholars were obliged to even criticise their own discipline in the *Dictionary of Home Economics* (Arima 2004), stating that, 'philosophical studies of home economics, whose role is to give an integrated consistency to further diversify home economics studies, have been cast into very difficult situations' (p.25).

The difficulty facing home economics' scientific identity is also reflected in its education in schools. As was the case with agricultural education, the post-war home economics education was reborn as a pedagogy for developing 'democratic' family members, which was designed to penetrate all its fragmented themes, including family meals. However, the term 'democratic family' soon vanished from textbooks in the mid-1950s to be replaced by more conservative (Confucianist) terms. This de-democratising tendency in home economics education was strengthened under economic development in which a 'good wife and wise mother' was viewed as convenient for male workers (Honoki 1990). More seriously, the presence of home economics education itself has continued to be weakened since the 2000s by reducing the required number of classes in primary schools, both absolutely and relatively, in favour of relaxed (*yutori*) education and other technical studies (e.g., computer studies).

The individual themes are certainly not free from this general tendency. In the post-war home economics education, the theme 'food' has been gradually expanded, while remaining de-politicised, to cover more nutritional and technical issues, and to establish, finally, three core objectives of teaching 'what' (e.g., nutrients and their bodily functions), 'how much' (e.g., required intakes) and 'in what ways' (e.g., cooking and table manners) to eat (Sato et al. 2014). Nevertheless, there has been no systematic perspective about how teaching these concrete themes can contribute to the construction of a 'democratic family' as its initial objective.

Next, I briefly turn to the historical transition of school meals (Fujihara 2018). During the pre-war period, pioneering activities to provide school meals gradually emerged mainly for children in need, such as the poor and those affected by natural disasters. Under the post-war American occupation, school meals started to be institutionalised at the national level to respond to the desperate food shortage facing children, which led to the School Lunch Programme Act in 1954. In this process of institutionalisation, both the authorities and practitioners developed a consensus on the total educational impacts of school meals, extending from the promotion of nutritional knowledge, the correction of picky eating, the valorisation of local products, and the development of 'democratic subjects' (although the democracy's meaning was often mistakenly reduced to the equal treatment of all children regardless of socio-economic background). Since the late 1960s, school meals have undergone an increasing neoliberal rationalisation; for example, by introducing the central kitchen system and favouring industrial (frozen, standardised and cheap) products, while its status as 'lively educational material' was, however paradoxically, upgraded 'from the peripheral to the central part of education' under the contemporary food education policy (p.254).

1.4.2 Research Programmatic Nature

This brief sketch of the histories of home economics and the school lunchtime programme, which now serve as the major frameworks for Food-related Guidance in schools, informs us of their programmatic nature. These pedagogies are premised on the basic value of 'good eaters who contribute to constructing a democratic family' through shopping, cooking and consumption. However, home economics as these pedagogies' disciplinary backbone has been unable to provide a genuine social theory for constructing such a 'democratic family' and has been reduced to nutritional and technical learning about eating, which can be eventually almost equated with nutrition education. Reflecting this ontological difficulty, home economics education has gradually shifted its practice space from the specific curriculum (whose number of classes is on the decline) to the school lunch programme, in which, however, home economics needs to adopt a new but difficult role, such as, resisting neoliberal rationalisation and responding to the widening socio-economic gap between children. The quasi-nutritional theoretical model of home economics education seems insufficient for addressing these challenges.

Despite the differing socio-political contexts, this programmatic nature is to some extent relevant to other countries (such as the UK and the US), in which the home economics has experienced a disciplinary identity crisis (Bennett 1978; National Institute for Educational Policy Research 2005). The school meal system has been positioned as an effective approach to addressing child poverty but increasingly challenged in the neoliberal political climate (Yada 1994; Levine 2008; Evans & Harper 2009).

1.5 The Basic Law on Food Education and Afterwards

1.5.1 The Birth of Food Education Policy

In 2005, the Basic Law on Food Education was enacted, and *shokuiku* finally became a socially and institutionally recognisable matter. All the existing food pedagogies (nutrition, agricultural and home economics pedagogies) were mobilised under this political framework. However, it is worth asking how the different principles of these precursor research programmes were integrated into this new concept, and in this regard, I briefly trace the process of its policy development.

At the political level, the term *shokuiku* had been used since 2002, when two important governmental documents were issued, the *Report of Research Committee on BSE Issues* and the *Plan for the Recovery of Food and Agriculture*, in which food education was positioned as a method of 'risk communication' or 'ensuring food safety and security of the consumers'. This positioning is said to result from the proposition by a journalist who was consulted by the Research Committee on BSE Issues and who claimed that risk management requires consumers' 'self-responsibility and self-decision' and 'it is food education that serves as such risk communication' (Nakamura 2012). Note that food safety management is primarily a matter for administrators and business operators, not consumers, and that risk communication is 'the exchange of information and opinions concerning risk and risk-related factors among risk assessors, risk managers, consumers and other interested parties' (FAO & WHO 2018) and, therefore, cannot be equated with food education.

The journalist's proposal was uncritically accepted by the working group of the Liberal Democratic Party, and the final draft (2004) was prepared in relation to the relevant ministerial objectives (agriculture, health and education). In this process, criticism from opposing parties concerning, for example, 'the possible violation of individual freedom of food choice' and 'the conceptual ambiguity of food education', went examined (Morita 2004), and the implications of the proposal have been overlooked by food education scholars, too. It is hardly possible to say that the principles of precursor food pedagogies were scrutinised and critically integrated into current food education policy, despite its acclaimed 'total' and 'comprehensive' approach to food.

1.5.2 The Post-Shokuiku Situation and International Perspective

To put this political framework into practice, the Committee on Food Education, comprising ministers and experts, was established (Article 16). This state body has developed the National Plan for Promoting Food Education every five years, based on which each local authority also has to develop plans according to territorial specificity (Art. 17–18). To trace the transition of political scopes, the First National Plan (2006–2010) articulated nine measurement indicators for evaluating its political impacts. The Second National Plan (2011–2015) presented three priority objectives: life stage-specific food education, prevention of lifestyle-related diseases and children's commensality with family members. The Third National Plan (2016–2020) extended its priority to the improvement of food environments and the transmission of food cultures (which responded to UNESCO's inscription of *washoku* as intangible cultural heritage in 2013). The current Fourth National Plan (2021–2025) reorganised its wide-ranging priorities into three categories of 'health', 'sustainability' and 'digitalisation' (the last of which was undoubtedly the consequence of the global pandemic). So far, it is enough for readers to understand the broad scope of food education and the difference from the traditional nutrition education policy. These National Plans have also been appropriated by relevant ministries to generate force for both their existing and new policies, most notably educational policies such as the deployment of nutrition teachers (who are registered dietitians and, at the same time, accredited teachers) in each school (2005) and the articulation of 'food education' in the national curricular guidelines (2008).

In addition, comprehensive research on international nutrition and food education policies were conducted as stipulated in the Basic Law's Article 25 (Cabinet Office of Japan 2007, 2008), including the US, the UK, Australia, France, Italy, Germany, the Netherlands, Denmark, Norway, Sweden, Singapore and South Korea. Although already discussed in the introductory chapter, again, I briefly comment on two general characteristics of this international political movement. First, no country prepared a comprehensive legal framework (not just an individual policy) dedicated to food education before 2008. However, France has increasingly institutionalised its food education policy since the mid-2010s with much greater established theoretical support than Japan, and their success has continued to inspire the other Western countries (Italy, Denmark, Finland, etc.). On the other side of the globe, Japan's early establishment of a food education policy encouraged East Asian countries, such as South Korea's Dietary Life Education Support Act in 2009, Taiwan's Food and Agriculture Education Law in 2022, which stipulated the implementation of school-based food education, and the experimental food education trials in Shanghai and Zhejiang in China (Pi & Akamatsu 2014). Second, the political frameworks in most countries before 2008 focused primarily on health promotion (notably the prevention of child obesity), not on the totality of food, which includes food systems, food cultures and food ethics. In addition to Japan, exceptions include food education policies in France and Italy, which contained a gastronomic focus on food cultures and taste education. Therefore, analyses of these three countries (though with greater emphasis on Japan and France) constitute the main contents of the following chapters.

1.6 Concluding Remarks

Taking the Japanese Basic Law on Food Education as a primary example, this chapter overviewed the historical context of how food education was constructed as an urging social and scientific theme. Applying a scientific-historical approach helped us to understand the research programmatic nature of the precursor food pedagogies and their complex dynamisms and interactions.

As the most dominant research programme, nutrition education is aimed at expanding one's ability for 'healthy eating'. The programme's theoretical basis primarily consists of nutritional sciences and social psychology and has been gradually expanded to meet contemporary demands since the 1970s (e.g., from health to well-being, from individual to socio-environmental factors); however, there is an increasing awareness of the limits of such bio-psychological theoretical ranges. Agricultural education generally takes the forms of a school garden programme and hands-on farming experience to develop democratic agents for 'social change in agricultural and rural environments'. Nevertheless, the programme's theoretical backbone has not been established, and thus, has been vulnerable to opportunistic political demands. Home economics education is aimed at constructing the 'democratic family' through teaching how to shop, cook and consume; however, it has suffered from similar problems to the nutrition and agricultural pedagogical programmes.

Each research programme has its unique hard-core value regarding 'good eaters' and its theoretical justification. From the perspective of scientific history (Lakatos, Berthelot and Laudan), it is inutile to criticise merely to refute this hard core. Instead, a more constructive approach is demanded to establish an integrative ontological and epistemological framework that can explain the scopes of these pedagogies and, thus, define 'food education' as a new rival research programme. In this regard, two implications can be obtained from the historical narrative in this chapter. First, despite the diversity in their educational scopes, these food pedagogies value particular aspects of eating (whether health, rural relations or family cooking), but not the 'totality' of eating. Second, each pedagogy shares the mission of developing children as 'democratic agents', although the pedagogy's theoretical basis remains de-politicised or simply unelaborated. Therefore, addressing these two elements should be one of the main tasks in constructing a rival research programme. Fortunately or unfortunately, the current policy was enacted without reflecting such historical and disciplinary complexities. This work hopes to revitalise the policy with valid theories and pedagogies.

Notes

1 As an exception, Takayama (1981) notes agricultural education by William Smith Clark (1826–1886), who was appointed by the Japanese Government from the Massachusetts Agricultural College in 1876 as a foreign advisor to establish the former Hokkaido University and who integrated liberal arts into agricultural education to 'cultivate the spirit (*shinden*) of man'.

2 Therefore, despite the historical and theoretical differences, nutrition education and agricultural education can also be performed within this 'Food-related Guidance' framework.

References

Bennett, R. (1978) What is home economics? *Journal of Consumer Studies and Home Economics*, 2, 79–84.

Berthelot, M. (2001) *Epistémologie des Sciences Sociales*. Presses Universitaires de France.

Brillat-Savarin, J.A. (1825) *Physiologie du goût*. Edited by J.F. Revel (1982). Flammarion.

Cabinet Office of Japan (2007) *Research Report on Food Education Policies in Other Countries*. https://warp.da.ndl.go.jp/info:ndljp/pid/9929094/www8.cao.go.jp/syokuiku/more/research/foreign/h19-1/index.html (accessed in August 2022).

——— (2008) *Research Report on Food Education Programmes in Other Countries*. https://warp.da.ndl.go.jp/info:ndljp/pid/9929094/www8.cao.go.jp/syokuiku/more/research/foreign/h20-1/index.html (accessed in August 2022).

Contento, I.R. (2010). *Nutrition Education: Linking Research, Theory, and Practice* (2nd ed.). Jones & Bartlett Learning.

Evans, C.E.L. & Harper, C.E. (2009) A history and review of school meal standards in the UK. *Journal of Human Nutrition and Dietetics*, 22, 89–99.

Farag, A.A., Badawi, S., Lalli, G. & Kamareddine, M. (2021) *School Farms: Feeding and Educating Children*. Routledge.

Food and Agriculture Organization of the United Nations & World Health Organization (2018) *Risk Communication Applied to Food Safety: Handbook*. FAO & WHO.

Fujihara, T. (2018) *Kyushoku no Rekishi [The History of School Meals]*. Iwanami Shoten.

Fujii, Y. (2014) Ishizuka Sagen's way of cultivating a whole person through chemical dietetics. *Bulletin of Faculty of Human Life Sciences of Fuji Women's University*, 51, 25–38.

Honoki, K. (1990) Minshuteki katei kensetsu to kateika [The construction of democratic families and home economics education]. In S. Iwadare & Fukuda, K. (eds.) *Kasei Kyoikugaku [Home Economics Science]*, 39–52. Fukumura Shuppan.

Ishida, A. (2000) The modern image of the consumer according to a novelist Murai Gensai in the Meiji era. *Kyoto Journal of Sociology*, 8, 31–50.

Ishizuka, S. (1898) *Shokumotsu Yojyoho [A Theory of Food for Health]*. Hakubunkan.

Kataoka, M. (2006) The development of the 'food, agriculture and environment' sides in agricultural education. *Japanese Journal of Agricultural Education*, 37(2), 69–80.

Kimoto, N. (2005) The process of institutionalisation of home economics in Japan. *Japanese Journal of Higher Education Research*, 8, 205–224.

Kuhn, T. (1962) *The Structure of Scientific Revolutions*. University of Chicago Press.

Kuroiwa, H. (2005) Wasurerareta meiiji no keimo shosetsuka [The forgotten Enlightenment novelist of the Meiji period]. In H. Kuroiwa (ed.) *Shokudoraku [Gourmandise]*, 577–589. Iwanami Shoten.

——— (2007) *Shokuiku no Susume [Recommendation of Food Education]*. Bungei Shunju.

Lakatos, I. (1978) *The Methodology of Scientific Research Programmes*. Cambridge University Press.

Laudan, L. (1977) *Progress and its Problems: Toward a Theory of Scientific Growth*. University of California Press.

Levine, S. (2008) *School Lunch Politics: The Surprising History of America's Favorite Welfare Program*. Princeton University Press.

Mah, C.L. (2010) Shokuiku: Governing food and public health in contemporary Japan. *Journal of Sociology*, 46(4), 393–412.

Ministry of Education, Culture, Sports, Science and Technology (2010) *Shoku ni kansuru Shido no Tebiki (revised edition) [Manual for 'Food-related Guidance']*. MEXT.

Morita, M. (2004) Shokuiku no Haikei to Keii [The background of food education in relation to the Basic Law on Food Education]. *Research and Information*, 457, 1–10.

Murai, G. (1903) *Shokudoraku (vol. 1) [Gourmandise]*. Edited by H. Kuroiwa (2005). Iwanami Shoten.

Murase, S. (2000) Shoku wo doraku ni suru hoho [The manual for making food as gourmandise]. In A. Kaneko, Takahashi, S. & Yoshida, M. (eds.) *Discourse no Teikoku [The Empire of Discourse]*, 167–198. Shinyosha.

Nakamura, M. (2012) *Symbol Kozo to Shugokoi wo meguru Dynamics: Shokuiku to Slow Food undo [The Dynamics of Symbol Structure and Collective Actions: Evidence from Food Education and Slow Food Movement]*. Sairyusha.

National Institute for Educational Policy Research (2005) *Kateika no Curriculum no Laizen ni kansuru Kenkyu: America [Research on the Improvement of Home Economics Educational Curriculum: The United States]*. National Institute for Educational Policy Research, 1–32.

Pi, W. & Akamatsu, J. (2014) Proposal for food education in China: By compare with the food education in Japan. *Bulletin of Center of Educational Research and Training of Wakayama University*, 24, 85–93.
Sato, M. (2009) Shokuiku movement of the Meiji era. *Hokkai-Gakuen University Journal of Economics*, 57(3), 87–96.
Sato, M., Kaneko, K. & Udaka, J. (2014) Changes to the description of food and nutrition in home economics textbooks of elementary and junior high schools in Japan. *Journal of Home Economics of Japan*, 65(10), 555–567.
Shimazono, N. (1989) *Eiyogaku no Rekishi [The History of Nutritional Science]*. Asakura Shoten.
Takayama, A. (1981) *Nihon Nogyo Kyoikushi [The History of Agricultural Education in Japan]*. Rural Culture Association.
Takemi, Y. (2002) Participatory nutritional education in the community: A framework for its evaluation. *Japanese Journal of Nutrition*, 60(2), 63–74.
Ueno, T. (2014) Nihon ni okeru Nogyosha Kyoiku [Agricultural education in Japan]. *Cooperative Finance for Agriculture, Forestry and Fishery*, 4, 29–31.
Yada, S. (1994) A study of school (meals) catering services in Great Britain. *Journal of Suzuka Junior College*, 14, 67–94.
Yamada, I. (2016) *Nogyo Taiken Gakushu no Jisho Bunseki [Empirical Analysis of Farming Activities]*. Association of Agriculture and Forestry Statistics.
Yamamoto, A. (1982) Gensai Murai's theory on dietary improvement. *Bulletin of Bunkyo Women's University*, 26, 44–56.

2 Research Structure and Epistemological Obstacles

Food education is envisaged as an 'interdisciplinary' or 'total' approach to food. From the historical perspective, the previous chapter questioned the actuality of this thematisation, and this chapter extends this point through in-depth analysis of the research structure of current food education. A systematic review was conducted of the eligible research articles (n=325) including the key term 'food education' (see Appendix for details of the methodology). Note that the scope of this review is strategically limited to the studies on food education in Japan to avoid mixing the discussion on each country's food education research in different socio-political conditions. Nevertheless, I later argue that the identified research structure can effectively explain the literature outside Japan. Drawing on these findings, I conclude this chapter by identifying two of the most fundamental epistemological obstacles in current food education, regarding its vey components – *food* and *education*.[1]

Before analysing the achievements and limitations of each discipline, it is useful to understand the overall structure of food education and relevant disciplines. As described in Figure 2.1, nutritional sciences, in which home economics (human ecology) is inseparably connected with, and thus, included, occupied 48.9% of the literature. The nutritional sciences also envisage the 'interdisciplinary' approach, but, in reality, merely refer to different approaches (home economics, cooking sciences, home economics education, etc.) within this discipline, not across disciplines. The medical sciences (including social medicine, public health and odontology) amounted to approximately 17.8% of the total literature. These two disciplines are inextricably linked with each other, and most of the literature was written by researchers in universities that offer training courses for dietitians and nutritionists. In other words, despite its 'interdisciplinary' reputation, the nutritional sciences constituted about 70% of the total literature, and thus, creates a general (and nutritionistic) understanding of current food education.

Diverse disciplines in the social sciences shared the other 30% of the literature. Of these, educational sciences have displayed an increasing interest in food issues but constituted only 5.2% of the literature. Moreover, 5.5% of the studies were conducted in the other social sciences (sociology, anthropology, history and business administration). Despite this minority position, this literature carries important criticisms regarding food education (which are reviewed soon later).

DOI: 10.4324/9781003341895-3

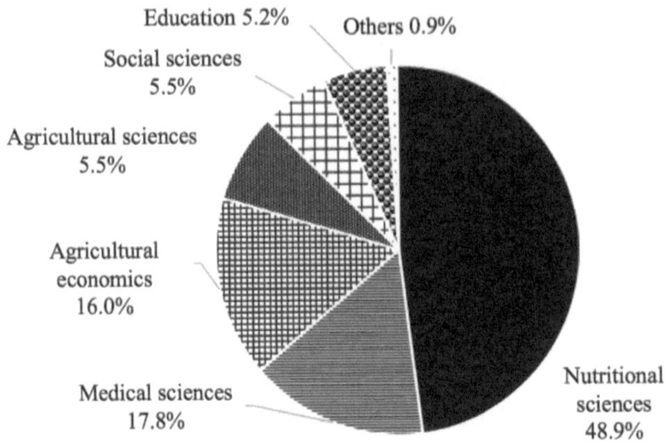

Figure 2.1 The Disciplinary Distribution of Food Education Literature
Source: Reprinted with permission from Ueda (2019, p.132).

However, most of the literature was written in English by international scholars located outside Japan and, therefore, has been poorly referenced in the articles in Japanese, which clearly reflects the *sakoku* (isolationist) characteristics of food education research in Japan.

In the social sciences, agricultural economics (otherwise called rural sociology) was the most dominant single discipline, constituting 16.0% of the literature, and it presented the most wide-ranging points of analysis. Therefore, the next section begins by reviewing this discipline.

2.1 Three Analytical Themes in Food Education Research

2.1.1 *Conceptual Studies*

Three analytical themes were identified from the literature review on agricultural economics (n=54). The first domain is aimed at defining the educational objectives and contents of food education and critically investigating the negative consequences of its conceptual ambiguity. In this domain, scholars have identified the following three conceptual problems.

The first critical perspective concerns current food education's 'nutrition-centred' approach and the resulting neglect of other socio-cultural values of food. For example, Ikegami (2008) conducted an in-depth analysis of university training courses for dietitians, who are supposed to be the major actors of food education, and problematised the absence of courses that dealt with the total values of food. In a different vein, Sato (2004) analysed the historical transition of themes of 'social studies' in primary education and pointed to the decreasing focus on agriculture and food systems.

The second conceptual problem concerns the distinction from existing food pedagogies. As already detailed in Chapter 1, there have been different traditions (such as nutrition, agriculture and home economics education), and scholars have attempted to differentiate the histories and principles of each pedagogy. However, as mentioned by Kataoka (2006), the overarching political concept of 'food education' had already vacuumed up these pedagogies without characterising their pedagogical merits and weaknesses.

The third group of studies points to some ideological aspects of food education due to its undefined nature and resultant vulnerability to political interests. Ikegami (2008) highlights that contemporary food education idealises (i) 'neoliberal' subjects that can well adapt to the market logic, (ii) the 'patriarchal' family regime based on the sexual division of labour (i.e., food provision by women), and (iii) the 'nationalistic' direction in favour of traditional food cultures. Although important ideological aspects were identified here, the theoretical basis for such criticisms is unclear and should be strengthened, a part of which effort is reviewed in the section on other social sciences.

2.1.2 Studies on Institutionalisation

The second analytical theme concerns how to establish an effective and sustainable system for promoting food education, because most food education programmes tend to be transient practices. In accordance with the progress of the food education policy (since the enactment of the Basic Law on Food Education in 2005), the focus of analysis has shifted from the 'motivational' factors of potential actors to the 'institutional' factors for sustainable promotion. Several case studies (e.g., Kataoka et al. 2012) have reported creative strategies (such as constructing community gardens); however, a systematic understanding of such effective strategies is still missing.

School meals as an institution have attracted the most scientific and social attention in this group of studies, partly because the Basic Law clearly situated the promotion of local products and traditional recipes in school meals as one of the most important food pedagogies for children. A fair number of empirical studies (e.g., Naito 2010) have followed this movement and identified the structural problems to be solved, such as the neoliberal preference for central kitchen systems, the lack of cooperation between distributors and dietitians, and the centralised menu development, all of which reduce flexibility in introducing local products. On the other hand, the educational impacts that can be brought by these institutional changes should not be neglected. This final point motivates the following third group of studies.

2.1.3 Evaluation of Educational Impacts

Studies on the educational impacts of food education can be largely divided into those for individual educational programmes and those for food education policy in general. The former faced methodological problems such as the ad-hoc selection of measurement indicators and the lack of pre- and post-evaluations. In agricultural economics, only Ueda (2017) has developed a comprehensive evaluation system for taste education programmes, which is discussed in greater detail in Chapter 4.

The latter studies are still burgeoning, and some econometric analyses have been made regarding children's problematic eating behaviours by using the statistical data obtained under the food education policy. For example, Kaneko (2016) econometrically demonstrated that children's solo eating was caused by changes in their lifestyles (cram schools, after-school activities, etc.), not by mothers' social promotion and extension of working hours, which had been often problematised in nutritional sciences. This second type of studies can help us investigate whether there are any ideological effects (e.g., the gendered view in Kaneko's analysis) in current food education.

2.2 Ideological Nature of Food Education

2.2.1 Nutritionism

Of the three analytical themes, conceptual studies are the major approach in sociology, anthropology, history and business management. Compare with agricultural economics, these studies have identified the similar ideological nature of food education but a little more elaborate such criticism. Regarding the place of nutrition, Kimura (2011a) introduced the concept of 'nutritionism' (Scrinis 2008), which takes nutritional science as the foremost authority for assigning values and virtues to food. Although she rather pointed to the de-nutritionistic trend in Japan for analysing private food education certification programmes, it would be appropriate to distinguish this practical dimension of food education from its theoretical one. The latter dimension is still within the nutritionistic paradigm.

2.2.2 Neoliberalism

Kimura (2011a) also extended the discussion to the neoliberal 'subjectivities' of food education. From political economy perspectives (such as Guthman 2008), all forms of food education have been condemned as narrowly focusing on improved food literacy and, thus, producing uncritical subjects regarding social structures. In rightly responding to this paralysing criticism, Kimura proposes an alternative framework for food education that incorporates structural analysis and does not shy away from challenging economic and political systems.

Similarly, the neoliberal principle of 'self-responsibilisation' was also problematised. In her discourse analysis of food education policies, Mah (2010) discovered the same self-responsibilising logic of food education as public health policies that have had since the 1990s and warns of the risk of neglecting the distribution of necessary commodities for achieving one's health. This type of reading has been repeated by Assmann (2015) and Takeda et al. (2016).

From the business perspective, a series of empirical studies on food companies' food education activities (e.g., Takano 2012) has highlighted some positive effects, including the improvement of corporate image and introducing creative educational methods; however, its negative consequences (such as making children docile consumers) have yet to be seriously explored.

2.2.3 *Gendered Ideology*

The principle of self-responsibilisation is directed not only towards individuals, but also at families, in which gender consciousness became an issue. In the abovementioned analysis of certification programmes, Kimura (2011a) identified the participants' gendered norms about women's roles in preparing family meals and warns of the risk of food education that reinforces this mentality. Moreover, Kimura (2011b) performed discourse analysis of government documents and elucidated that it is not the absence of fathers but the declining role of mothers as meal providers that was considered the problem in the food education discourse. Similarly, through interviews with Japanese citizens, Takeda et al. (2017) point to gendered differences in perceiving 'food education' and the consequential exclusion of men from the realm of food education. All these studies share the need for a gender-inclusive approach in food education.

2.2.4 *Nationalism*

The final ideological dimension concerns nationalism due to the uncritical idealisation of 'traditional Japanese diets' or simply 'Japanese diets'. Based on her discourse analysis, Takeda (2008) clarifies how only Japanese elements (such as rice) were normalised as 'delicious food in a beautiful country' despite the hybridity inherent in the historical development of Japanese food culture. Using the Foucauldian perspective (Foucault 1976; Coveney 2010), she concludes that the banal nationalism performed in daily food education functioned as a 'biopolitical device' that links individuals with a multitude of governing bodies, including the nation-state. The discipline of history was also echoed with the *sakoku* (isolationist) perspective that current food education emphasised – only the uniqueness and uniformity of Japanese food culture – and made subjects ignorant of its historical interaction with foreign cultures (Fujihara 2016).

2.2.5 *Theoretical Basis and Its Limits*

These criticisms above were informed by Foucault's theory of power, Althusser's (1970) theory of ideology, and political economy (a particular scholarship in the US such as Guthman 2008). Each criticism carries an element of the truth and helps to reinforce the theoretical backbone of the conceptual studies. However, one should be careful about this criticism's limits regarding data availability and the universality of implications. As noted above, most of these critical scholars are located outside Japan and, perhaps due to this geographical gravitation, the main source of their discourse analysis was confined mainly to governmental documents (the Basic Law, conference minutes, etc.). These sources could provide the logic of state-run food education but not the entire scope of food education practices involving academics, practitioners and ordinary individuals and families. As analysed by Foucault in his genealogical analysis of madness, sex and humans, discourse analysis should be based not only on government documents, but also on much more diverse, rich and nuanced sources. These criticisms' implications about the ideological nature of food education are insightful but should be interpreted with some caution.

2.2.6 Food Pedagogies and Inquiry into Ethics

In the educational sciences, particular importance can be given to the special issue on food education in the *Australian Journal of Adult Learning*, later published as *Food Pedagogies* (Flowers & Swan 2018). This issue covered food education in several countries, including the analysis of Japan's food education (Reiher 2012). In contrast to the prevailing use of 'food education', this issue presents the analytical concept of 'food pedagogies' from the perspective of educational sociology. In this sense, food education can be conceived as all types of pedagogical processes inside and outside classrooms (including television and socialisation at daily tables) in which various food technologies (school curricula, food labels, etc.) and power relations in such pedagogical processes become the major point of analysis.

Interestingly, such power analysis has yet to be adopted by Japanese educationalists on food education. Rather, analytical priority has been given to a more normative approach, perhaps because such a socially constructive approach would risk further complicating the already-ambiguous nature of food education in Japan. Although this type of normative study is still burgeoning, Shinmura (2015) points to the increasing ambiguity of what constitutes 'human development' in food education policies and proposes to articulate this ideal through the historical analysis of educational philosophies from antiquity until today (Plato, Aristotle, Rousseau, Dewey, etc.).

2.3 Achievements and Limitations in Nutritional Sciences

2.3.1 Evidence-Based Approach Without Conceptual Exploration

Keeping these findings in mind, I finally turn to the thematisation of food education in the nutrition sciences. I have discussed how food education research comprises three major analytical themes: concepts and ideologies, institutionalisation and educational impacts. Of the total literature in this discipline (n=205), almost 80% (n=155) were categorised as studies on educational impacts, while those on institutionalisation (n=26) and educational impacts (n=24) constituted about 10% each.

Before scrutinising the first group of studies, I must comment on the (almost) complete absence of conceptual exploration of food education, despite its disciplinary dominance occupying about 70% of the literature. I identified only one article exploring 'how food education should be' (Adachi & Eto 2005), in which food education is defined as 'the effort for better and more harmonious coexistence between people's quality of life (QOL) and quality of environment (QOE) by educating on the whole aspects of food and nutrition history, building dietary capability and creating a practical nutritional environment' (p.1). This definition was derived from the Japanese scholarship of 'food ecology' (somewhat similar to the ecological model in the American nutrition education); however, the validity of its theoretical basis has yet to be scrutinised by global scholarship. In addition, note that the terminologies here, such as 'QOL', 'whole aspects of food' and 'capability', are non-theoretical, and thus, cannot be equated with those in the CA or sociology of food pursued in this book. Despite its unelaborated concept, this definition has been popularised in the following conceptual studies, none of which critically challenge it.

Nevertheless, the major achievements in this discipline can be identified in the studies on educational effects on children. These studies can be largely divided into two subgroups: research on children's dietary behaviours prior to effective programme developments (n=101), and those on the intervention and evaluation (n=54). The latter studies were reviewed regarding their multiple constituents: (i) *Educational targets*: most studies were targeted at food education programmes for elementary school children within multiple subjects (home economics, integrated study, extracurricular activity, etc.); (ii) *Evaluation design*: only about 30% of the studies employed non-randomised control trials, whereas pre- and post-evaluation or only post-evaluation were used in the other 70%. The dominance of case studies with relatively low evidence implies the difficulty of effectively comparing results and articulating the general educational impacts of food education; (iii) *Evaluation methods*: qualitative measurements and questionnaire sheets were two major evaluation methods. The former included elements such as BMI, food intake and a sensory test, whereas the latter covered a wide range of indicators (knowledge, attitude and behaviours) and often included unvalidated or too simplified indicators for educationally important but difficult-to-measure qualities such as 'food curiosity' and 'pleasure of eating'; (iv) *Educational contents*: although seemingly diverse, certain established groups of educational contents could be identified. The first group concerns the specific meal composition of 'staple, main dish and side dish(es)', often normalised in food education policies. However, this aspect does not mean the presence of systematic pedagogies for enabling children to eat such a meal composition. The lack of 'systematicity' is also problematic regarding all other educational content, except for only a few studies on 'taste education' and food ecology-based pedagogies (such as the 'bento method' and 'whole fish programme').

2.3.2 *Possible Answers to Critical Studies*

As noted, various criticisms about the ideological nature of food education were not acknowledged in the nutritional sciences, and the absence of conceptual exploration was identified as one of these reasons for this. On behalf of nutritional scientists, I offer possible answers to these criticisms by drawing on the findings.

First, although a wide range of educational content (meal content, taste, cooking, farming, etc.) seems to be proposed in current programmes, closer inspection reveals that most of these programmes justified their social themes on the nutritional account. The meal composition of 'staple, main dish and side dish(es)', for example, was valued uniquely for its health aspects, rather than the historical or food system accounts. Similarly, the important elements of Japanese food cultures, such as miso and green tea, were valorised in terms of their richness in proteins and polyphenols, respectively. This point confirms the nutritionistic paradigm (Scrinis 2008) is deeply rooted in food education developed in nutritional sciences.

Second, the nationalistic tendency was not very salient in these disciplines. This aspect was perhaps the other side of educational programmes being nutrition-centred, and thus, not including cultural analysis. However, the critical perspective (as proposed in the sociological literature) is still needed because some programmes dealing with *washoku* were naïvely falling within the nationalistic paradigm. Moreover, as Fujihara (2016) points out as a concern, only a few programmes included foreign food cultures in their educational content.

Third, most educational effects on children were generally evaluated by gender, but a more in-depth analysis regarding gendered norms internalised in daily eating could not be identified. More seriously, no study integrated class (income, education or occupation) indicators of families. This lack highlights a need to develop a theoretical framework that can situate the diversity of situations facing children (later theorised using the CA).

Finally, the neoliberal criticism also rightly reflected the actuality in nutritional sciences. Although some programmes included school excursions to rural farms or hands-on experience, the intended educational impacts were too psychological (such as 'a sense of gratitude to nature and farmers') and too uncritical regarding the structural problems facing rural societies and agriculture (e.g., super-ageing rural demographics, a significant decline in the number of agricultural workers, unfair food systems in which farmers cannot cover even their production costs).

2.4 Japan's Relevance to International Literature

I have so far articulated the research structure of food education as characterised by three major analytical themes, the dominance of nutritional sciences and its disciplinary (de)connections. Although the focus of the review was limited to food education in Japan, a certain universality of this structure can be assumed for food education in other countries. The identified structure helps to position a growing body of international publications on food education in the disciplinary map. Let me briefly illustrate this point.

In many countries, food education has been traditionally addressed as 'nutrition education', and numerous textbooks have been published primarily aimed at to-be-dietitian/nutritionist university students (most notably Contento 2010). These textbooks present orthodox theories of nutritional sciences and social psychology to develop food education programmes effectively, but the fundamental problems facing these scholars are the same as those of Japanese food education in nutritional sciences (discussed further soon later).

Critical studies (feminist studies, political economy and sociology) constitute another strain of increasing literature (such as Hayes-Conroy 2014; Flowers & Swan 2018; Wistoft & Qvortrup 2019). All these works have helped to highlight the socially constructed nature of food education, but they largely share the same underlying motivations (i.e., its ideological effects) and theoretical bases (Foucault, Bourdieu and other postmodernist scholars) as those already addressed in the sociological section. Interestingly, some studies have displayed a particular interest in 'taste education', and these studies' implications and limits are discussed throughout this book (notably, Chapter 8).

The other strand of increasing literature can be observed in the field of educational sciences, which do not necessarily take the abovementioned critical and postmodernist direction (Hart 2016; Rice & Rud 2018; Lalli 2019; Hart & Page 2020). This strand is suggestive of the food education research in Japan, in which the educational studies account for only 5.2% of the literature, particularly in terms of the exploratory application of the capability approach (Hart 2016; Hart & Page 2020) and the Deweyan philosophy (Rice & Rud 2018) to food. Nevertheless, I have identified a gap in the knowledge between this group of 'educational' studies and the much wider range of 'food' studies. Therefore, this book should fulfil the task of addressing this missing link.

Although in this chapter the research structure was analysed mainly from the disciplinary perspective, there are other possible approaches. From the geographical perspective, extensive studies have emerged on food education in the UK. (Hart 2016; Tull 2018; Lalli 2019), the US. (Hayes-Conroy 2014; Rice & Rud 2018) and Denmark (Wistoft & Qvortrup 2019), as well as many fragmented studies on other countries. In this sense, this book is the first study to provide a comprehensive picture of food education in Japan and France, two countries of political and theoretical importance. From the pedagogical perspective, there have been studies on nutrition education (Contento 2010), agricultural education (Farag et al. 2021) and home economics education (Tull 2018; Rutland & Turner 2020), with different terminologies in each country. However, the challenge is the lack of any conceptual and theoretical frameworks that can effectively explain the rich diversity of food pedagogies, which is also addressed from the capability perspective in this book.

2.5 Concluding Remarks: Epistemological Obstacles

Drawing on these findings, I conclude this chapter by summarising two of the most fundamental epistemological obstacles in current food education, regarding its very components – *food* and *education*. First, despite food education's envisaged 'interdisciplinary' scope, almost 70% of the literature was produced within the nutritional sciences. Although critical studies in social sciences have highlighted the ideological (neoliberal, gendered, nationalistic) nature of food education, these criticisms were not integrated into the current dominant paradigm, which left its 'nutritionism' (Scrinis 2008) unaddressed and the total values of food reduced to a single nutritional factor. To overcome this nutritionism, from the scientific history perspective (Chapter 1), it is essential to construct a new rival research programme (Lakatos 1978) with improved explanatory power for food education, rather than merely proposing 'critical' approaches. The following chapters begin with the diagnosis of taste education as a promising alternative for learning the total (bio-psycho-social) values of food in terms of its pedagogical nature (Chapter 3), educational effects (Chapter 4) and institutionalisation (Chapter 5), which correspond with the three major analytical themes of food education research identified in this review.

Second, no established educational or ethical theory was proposed in the previous literature, and thus, the ethical nature of food education has been unaddressed in food education research. Critical pedagogical theories have been employed, but their main tasks were to deconstruct the norms regarding 'how humans should be/eat', rather than to construct and inquire about new norms. Although traditional nutrition education has strengthened its focus on the participation and empowerment of subjects in the educational process (as seen in Chapter 1), surprisingly, no established educational theory was included in Japanese textbooks, nor in their original versions in the US (Glanz et al. 2008; Contento 2010), except for occasional revisits to some educationalists (notably, John Dewey and Paulo Freire). What is at stake regarding current food education is not only nutrition and health, but also the ethical dimension of eating, particularly concerning eaters' well-being, freedom, equality, social justice and human development. In contemporary society, in which inquiries into 'virtue ethics (*telos*)' became absent (MacIntyre 1981), what is far more urging is a normative framework that helps us to investigate 'how humans should eat' and constructs the ethical foundation of food education. This problem is addressed practically by the in-depth analysis of taste education and gastronomy (Parts I–II) and theoretically by applying Amartya Sen's CA (Part III).

Note

1 The first three sections (Sections 2.1–2.3) were reprinted with permission from Ueda (2019) Achievements and challenges of food education research in agricultural economics and implications from relevant disciplines through interdisciplinary literature review. *Journal of Food System Research*, 26(3), 128–143.

References

Adachi, M. & Eto, K. (2005) Expectations for 'Shokuiku' (food and nutrition education and promotion). *Japanese Journal of Nutrition and Dietetics*, 63(4), 201–212.

Althusser, L. (1970) Idéologie et Appareils Idéologiques d'Etat (Notes pour une recherche). *La Pensée*, 151, 3–38.

Assmann, S. (2015) The remaking of a national cuisine: The food education campaign in Japan. In J. Farrer (ed.) *The Globalization of Asian Cuisines*, 165–185. Palgrave Macmillan.

Contento, I.R. (2010). *Nutrition Education: Linking Research, Theory, and Practice* (2nd ed.). Jones & Bartlett Learning.

Coveney, J. (2010) *Food, Morals and Meaning: The Pleasure and Anxiety of Eating*. Routledge.

Farag, A.A., Badawi, S., Lalli, G. & Kamareddine, M. (2021) *School Farms: Feeding and Educating Children*. Routledge.

Flowers, R. & Swan, E. (2018) *Food Pedagogies*. Routledge.

Foucault, M. (1976) *Histoire de la Sexualité (vol. 1): La Volonté de Savoir*. Gallimard.

Fujihara, T. (2016) Shoku wa kyoiku no kadai nanoka [Is food a matter of education?] In T. Sato (ed.) *Gakushu suru Shakai no Asu [The Future of Learning Society]*, 179–202. Iwanami Shoten.

Glanz, K., Rimer, B.K. & Viswanath, K. (2008) *Health Behavior and Health Education: Theory, Research, and Practice*. John Wiley & Sons.

Guthman, J. (2008) Neoliberalism and the making of food politics in California. *Geoforum*, 39(3), 1171–1183.

Hart, C. (2016) The school food plan and the social context of food education in schools. *Cambridge Journal of Education*, 46(2), 211–231.

Hart, C. & Page, A. (2020) The capability approach and school food education and culture in England. 'Gingerbread men ain't gonna get me very far'. *Cambridge Journal of Education*, 40(6), 673–693.

Hayes-Conroy, J. (2014) *Savoring Alternative Food: School Gardens, Healthy Eating and Visceral Difference*. Routledge.

Ikegami, K. (2008) Anshin anzen shakai ni okeru shokuiku no fuchi [The place of food education in a safe and secure society]. In K. Ikegami, Harayama, K. & Fujihara, T. (eds.) *Shoku no Kyodotai [The Community of Food]*, 177–237. Nakanishiya Shuppan.

Kaneko, J. (2016) With whom do children eat evening meal in Japan and why? Evidence from anonymous data of survey on time use and leisure activities in 2006. *Journal of Rural Problems*, 52(2), 23–33.

Kataoka, M. (2006) The development of the 'food, agriculture and environment' sides in agricultural education. *Japanese Journal of Agricultural Education*, 37(2), 69–80.

Kataoka, M., Takatsu, H. & Yamada, I. (2012) Factor for development and continuance of education farm of farmers group in Japan. *Japanese Journal of Farm Management*, 50(3), 72–77.

Kimura, A.H. (2011a) Food education as food literacy: Privatized and gendered food knowledge in contemporary Japan. *Agriculture and Human Values*, 28(4), 465–482.

——— (2011b) Nationalism, patriarchy, and moralism: The government-led food reform in contemporary Japan. *Food and Foodways*, 19(3), 201–227.

Lalli, G. (2019) *Schools, Food and Social Learning*. Routledge.

Lakatos, I. (1978) *The Methodology of Scientific Research Programmes*. Cambridge University Press.

MacIntyre, A. (1981) *After Virtue: A Study in Moral Theory*. Notre Dame Press.

Mah, C.L. (2010) Shokuiku: Governing food and public health in contemporary Japan. *Journal of Sociology*, 46(4), 393–412.

Naito, S. (2010) Gakko kyushoku ni okeru shokuzai chotatsu to zibasanbutsu no riyokakudai hosaku [The food supply for school meals and measures for facilitating the use of local products]. *Agriculture and Horticulture*, 85(2), 225–234.

Reiher, C. (2012) Food pedagogies in Japan: From the implementation of the Basic Law on Food Education to Fukushima. *Australian Journal of Adult Learning*, 52(3), 507–531.

Rice, S. & Rud, A.G. (2018) *Educational Dimensions of School Lunch: Critical Perspectives*. Palgrave Macmillan.

Rutland, M. & Turner, A. (2020) *Food Education and Food Technology in School Curricula: International Perspectives*. Springer.

Sato, Y. (2004) Nogyo wa shokuiku wo dou uketomeruka? [How does agriculture accept food education?]. *Quarterly Journal of Food, Agriculture and Social Studies*, 70(12), 67–79.

Scrinis, G. (2008) On the ideology of nutritionism. *Gastronomica*, 8(1), 39–48.

Shinmura, H. (2015) Gakko kyushoku shokuiku no konnichiteki igi to kadai nitsuite [On the contemporary significance and challenge of school meals and food education]. *Quarterly Education Law*, 186, 28–33.

Takano, K. (2012) Business opportunity and food education in Japan. *Social Responsibility Journal*, 9(4), 516–533.

Takeda, H. (2008) Delicious food in a beautiful country: Nationhood and nationalism in discourses on food in contemporary Japan. *Studies in Ethnicity and Nationalism*, 8(1), 5–30.

Takeda, W., Banwell, C. & Dixon, J. (2016) Advancing food sovereignty or nostalgia: The construction of Japanese diets in the national Shokuiku policy. *Anthropological Forum*, 26(3), 276–288.

Takeda, W., Kelby, M. & Ishikawa, Y. (2017) Food education for whom? Perceptions of food education and literacy among dietitians and laypeople in urban Japan. *Food Studies: An Interdisciplinary Journal*, 7(4), 49–66.

Tull, A. (2018) *Food and Cooking Skills Education: Why Teach People How to Cook?* Routledge.

Ueda, H. (2017) Impact evaluation of a school-based food education intervention on cognitive and attitudinal aspects of children's dietary habits: A case study of a sensory education program 'Leçon de Goût'. *Journal of Food System Research*, 24(1), 2–17.

——— (2019) Achievements and challenges of food education research in agricultural economics and implications from relevant disciplines through interdisciplinary literature review. *Journal of Food System Research*, 26(3), 128–143.

Wistoft, K. & Qvortrup, L. (2019) *Teaching Taste*. Common Ground Research Network.

Part I

Taste Education as an Alternative Pedagogy

3 History and Pedagogical Nature

In the previous chapter, 'pedagogical ambiguity' was identified as one of the major challenges in the current food education literature. This ambiguity not only makes food education vulnerable to ideological effects, but also complicates our understanding of its general educational impacts on children. Moreover, it hinders possible cooperation among food education actors. In light of this ambiguity, the focus of this chapter is on taste education, which is the most systematic pedagogy being promoted in France, Japan and Italy.[1]

Considering that taste education consists of a group of pedagogies with internal diversities and their complex interactions, the answer to the question of what taste education is cannot be regarded as self-evident. Therefore, I first analyse the genesis and transmission of several taste education models in the above three countries to elucidate their complex pedagogical interactions. Note that, in this chapter, I limit myself to briefly outlining their history. Further analysis is provided in Chapter 5. The data for analysis included the relevant literature (textbooks, research articles, etc.). If necessary, the leaders of the promoter organisation were contacted for further clarification.

Second, three of the most established methods in these countries were selected (France: Puisais; Italy: Slow Food; Japan: Shinagawa). The textbooks in which these methods are described are analysed to identify their pedagogical commonalities, with a particular focus on their educational objectives, theories, educational content and materials, and teaching methods. Generally, there are two types of methodological approaches to analysing textbooks: One is to interrogate its pedagogical effects on readers, whereas another is to inquire about its educational norms (Fujioka 1991). The latter approach is adopted in this chapter.

In this chapter, the analytical focus shifts to the institutional dimensions of taste education. As noted in Chapter 2, most food education programmes tend to be transient. Therefore, an institutional analysis, with a view to establishing its sustainable promotion, constitutes the second thematic group of food education research. Taste education is a promising case because it can bring a unique perspective grounded in nearly 50 years of experience, having been around since the 1970s. It also enables us to perform international comparative analysis (France, Italy and Japan) to create universal implications for a more sustainable institutionalisation. Thus, using health promotion theories in public health, each country's

DOI: 10.4324/9781003341895-5

system of promoting taste education is examined in this chapter, with a special focus on (i) its political positioning (ii), the organisational structures of promoter-actors and (iii) their effective strategies within institutions and communities.

3.1 Genesis and Transmission of Various Models

3.1.1 The Puisais Method and Its Derivatives

In France, it is necessary to distinguish the original Puisais method from its derivatives. A complete history of taste education is provided in Chapter 5; thus, in this chapter, I draw only its outline to highlight such pedagogical differences. Jacques Puisais (1927–2020) was a French philosopher, oenologist, poet and humanist of taste. He was also the founder of French taste education. The promoter institution (*Conservatoire du Goût*) was established in 1974. It offered its first programme of ten sessions for elementary school children in 1975 (Puisais 1999). An updated programme of 12 sessions was developed in 2002 (Puisais et al. 2002), the educational impact of which was proved by a national survey called 'EduSens' in 2005 (Schlich et al. 2014). An abbreviated programme of eight sessions was developed in 2011 (Montoux et al. 2012) and has since served as the recommended pedagogy of relevant ministries (notably, education and agriculture). A national network of taste education actors and a system of support from the local authorities were later constructed. Under the framework, two types of pedagogies – the original one that rigorously follows Puisais' method and other derivatives that modify it to their territorial contexts – have been practised and should be distinguished from each other. In another vein, the popularity of the Puisais method eventually led to the *La Journée du Goût* (The Day of Taste) initiative in 1999, in which 350 chefs provided taste education classes for children in Paris, and later *La Semaine du Goût* (LSG: The Week of Taste) from 1992 (Togawa 2009). The LSG method has been continuously practised under the national framework, but should also be differentiated from the original Puisais method.

The textbook for the first ten-session programme, *Le Goût chez l'Enfant* (Puisais 1999), best represents the pedagogical identity of Puisais' method. Moreover, it has been popularised (e.g., there are two translated versions in Japanese). It is, thus, this version that is scrutinised in the following textbook analysis.

3.1.2 Prato Method and Slow Food

The evolution of Italian taste education has been closely connected with that of the Slow Food movement. Although Carlo Petrini is perhaps most well-known for this movement, the Slow Food method actually originated from Alessandro Venturi, who was local historian and one of Slow Food's founding members, and his partner, Luisa Peris, who was an elementary school teacher. Through a series of trials in schools in the city of Prato, their unique method, which I hereafter call the 'Prato Method', was developed with the support of Slow Food Prato (which Venturi himself established in 1987). This method has been gradually institutionalised with the support of the Ministry of Education, a process that began in 1999. Venturi and Peris also contributed to establishing the educational division of Slow Food Italy in 1998 and later to developing the Slow Food method. The textbook for this method, *In What Sense? A Short guide to Sensory Education* (Barzanò & Fossi 2010), has been translated into multiple languages and popularised internationally. In contrast to the Slow Food movement, which increased its political orientation, the primary objective of Venturi and Peris remained the territorialised practice in Prato. They therefore decided to become independent from Slow Food Prato in 2007 and established the *Centro Educazione del Gusto* (CEG: Taste Education Centre), in which the original Prato method has been practised until now (CEG & Nakano 2012). Taking these institutional changes into account, it is plausible to distinguish the Prato method from its derivative (but much better known) Slow Food method. The latter is the target of textbook analysis.[2]

3.1.3 Hybridised Methods in Japan

These French and Italian methods were transmitted to Japan in the late 1990s and early 2000s, which generated at least five hybridised methods (one of which is relatively independent from foreign influences). The French LSG method was pioneered in Japan in 1998, when a cooking specialist, Yoshimi Uchisaka, based on her observation of the actual practice in France and its educational guidelines, taught the first class in Japan in which this method (*Leçon de Goût*) was taught to Japanese children. She later established the *Nihon Mikaku Kyokai* (Japan Taste Association) with prestigious chefs who prepared Japanese and French cuisine (Uchisaka 2007). Following this early introduction, two derivative pedagogies emerged. One was the Japanese LSG method, which has been increasingly utilised since 2010 by relevant ministries and private food companies, and justified with solid scientific evidence (such as Ueda 2017; Hayabuchi et al. 2017). The other derivative was the pedagogy developed by the MIIKU (the Japan Taste Education Association in Japanese) in 2015, whose founder had previously consulted Uchisaka and completed a modified (but largely different) method based on her own theory of cookery science (Miyagawa 2016). The original Puisais method was also transmitted to Japan, rather sporadically, by some participants in Puisais' own training programme in 2007 (e.g., Sato 2009) and, more institutionally, by the newly-established *Institut de Développement du Goût chez l'Enfant* (IDGE: Institute of Children's Taste Development) from 2012. The IDGE has rigorously followed the Puisais method, with a slight modification of tasting samples (the use of Japanese products) and the institute also published Puisais' translated textbooks (Ishii et al. 2016; Ishii & Tajiri 2017). Thus, at least three derivatives of the French tradition can be identified.

The Italian tradition, on the other hand, was transmitted to Japan together with the Slow Food movement. Prior to the establishment of its central organisation in Japan (Slow Food Japan) in 2004, some regional organisations had direct interactions with the Slow Food Italy headquarter and practised the Slow Food method in Japan. One of the representative cases is Slow Food Suginami Tokyo, which has utilised a unique method (which I hereafter call 'the Tokyo method') that is based on the Slow Food method, in extracurricular activities since 2002 (Sasaki 2008). Although it is certain that other actors have practised taste education techniques other than the Tokyo method, there is no literature at this time that will enable us to comprehensively understand their practices.

As a relatively independent movement, a unique Japanese pedagogy was also developed by a scholar of environmental education, Akira Shinagawa, who effectively integrated the ecological perspective and Japanese (or Eastern) thought into taste education. In the process of its institutionalisation, Shinagawa also had interactions with the abovementioned CEG in Italy, from which he received pedagogical inspiration. Shinagawa's method was first utilised in 2011 as a university course and has since been promoted by the Food Consciousness Institute (FCI) (Figure 3.1).

3.2 Pedagogical Nature of Taste Education

Each of the three textbooks is basically composed of (i) an introductory part, in which the method's background and educational objectives are explained; (ii) the main part, comprising the educational content; and (iii) the concluding part, in which the objectives are restated. (For the Slow Food method, this part is replaced

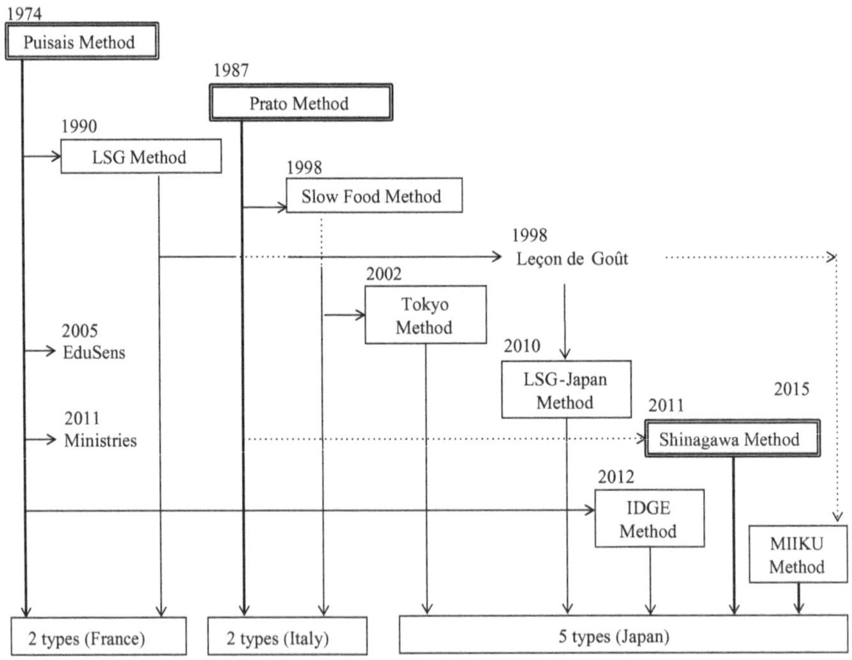

Figure 3.1 Pedagogical Interaction of Taste Education Methods in France, Italy and Japan
Source: Reprinted with permission from Ueda (2020, p.54). The establishment year refers to that of the promoter institution (not the pedagogy itself). The original pedagogies are boxed in double lines, while the variants are indicated by single lines. Direct transmission is indicated by solid arrows and partial interaction by dotted arrows.

with the corresponding cartoons.). Hereafter, I discuss each of the pedagogical qualities (Table 3.1).

3.2.1 Educational Objectives

The background and educational objective of the *Puisais* method were that

> we must give all the opportunities to children to choose gustative pleasures if we do not hope to make them the *handicapped of taste* [...] but how can we construct such pleasures of living at the table? To answer this question is all our objectives.
>
> (Puisais 1999, p.31)

Table 3.1 Structures and Contents of Textbooks

Pedagogy	*Puisais Method*	*Slow Food Method*	*Shinagawa Method*
Introduction	Preface (pp.7–11) Ch.1 Likes and Dislikes 1 Mechanism of Taste (pp.15–20) 2 Development of Taste (pp.21–26) 3 Diversity of Taste (pp. 27–31)	Ch.1 Introduction (pp.4–5) Ch.2 Method (pp.6–13)	How to Use the Textbook (pp.1–5) Introduction (pp.6–7) Ch.1 Objectives (pp.8–33)
Main	Ch.2 School of Taste 1 Introduction (pp.35–40) 2 Awakening Taste (pp.41–128) Session 1. Five Senses (pp.41–49) Session 2. Four Tastes (pp. 49–54) Session 3. Menu (pp.54–59) Session 4. Olfaction (pp.59–68) Session 5. Vision (pp.68–86) Session 6. Sense of Touch (pp.86–98) Session 7. Interference (pp.98–101) Session 8. My Region (pp.101–125) Session 9. Conclusion (pp.125–127) Session 10. Repose (pp.127–128)	Ch.3 Senses (pp.14–15) → S.1 Ch.4 Vision (pp.16–21) → S.5 Ch.5 Olfaction (pp. 22–28) → S.4 Ch.6 Sense of Touch (pp.29–34) →S.6 Ch.7 Taste (pp.35–44) →S.2 Ch.8 Hearing (pp. 44–47) → ◇ Ch.9 Be Conscious of Multiple Senses (1) Interference (pp. 48–51) → S.7 (2) Sensory Orchestra (pp.52–58) →S.9	Ch.2 Five Senses (pp.34–48) →S.1–2, 4–6, ◇ Ch.3 Taste (pp.49–69) →S.1–2, 4–6, ◇ Ch.4 Tasting (pp.70–82) →S.1–2, 4–-6, ◇, ◆(Mind) Ch.5 Deliciousness (pp.83–87) Ch.6 Japan and East (pp.88–95) Ch.7 Right of Taste (pp.96–100) Ch.8 Power of Taste (pp.101–103) Ch.9 Five Minds (pp. 104–105) → ◆

(*Continued*)

Pedagogy	*Puisais Method*	*Slow Food Method*	*Shinagawa Method*
Conclusion	(3) After... (pp.129–137) Ch.3 Practice at Family (pp.141–184) Conclusion (p.185) Acknowledgement (p.186) Bibliography (p.187)	Vocabulary for Taste (pp.59–61) Recommended Literature (pp.62–63) Cartoon for Child (pp.65–67) Cartoon for Adult (pp.68–70)	Ch.10 Education (pp.106–113) Ch.11 Structure (pp.114–116) Ch.12 Classroom (pp.117–121) Ch.13 Ingredients (pp.122–122) Conclusion (pp.123–124)
Background	'Handicap of Taste' – Commercialisation of food – Medicalisation of food	'Real Sensory Deprivation' – Globalisation of food – Industrialisation of food	'Impoverishment of the Five Tastes' – Globalisation of food – Delocalisation of food
Objective	– Reflection of one's own sense/body – Socialised pleasure	– Food choice of good qualities	– Personal development with Japanese/Eastern minds
Theory	– Physiology of taste – Child psychology	– Physiology of taste – Child psychology	– Physiology of taste – Japanese and Eastern thought (?)
Material	– Textbook – Tasting experience – Communication with others	– Textbook – Tasting experience – Communication with others	– Textbook – Tasting experience – Communication with others
Content	– Physiological laws of taste – Geography, history and food culture	– Physiological laws of taste – History and food culture	– Physiological laws of taste – Japanese view on nature
Teaching method	(1) Theory learning (2) Tasting (3) Communication with others	(1) Theory learning (2) Tasting (3) Communication with others	(1) Tasting (2) Communication with others (3) Theory learning

Note: The 'S' in the Slow Food and *Shinagawa* methods corresponds with the sessions in the *Puisais* Method.

Puisais identifies the excessive commercialisation and medicalisation of food (pp.30, 185) as the root causes of producing such handicapped children. Another key concept, 'pleasures', is later specified as being generated by 'being involved in a pleasant relationship with one's own body and others' (p.185). Such pleasures of eating are not materialistic ('obsessed with folks') and closed, but rather 'relational' and 'open' to all cultures (p.185). In other words, Puisais' method is defined as a pedagogy whose primary objective is to cultivate children's abilities to construct pleasurable relationships with their own bodies or senses, as well as with others (of other cultures).

Slow Food also justifies its existence by problematising children's 'real sensory deprivation' (Barzanò & Fossi 2010, p.4), which is akin to Puisais' concept of the 'handicapped of taste'. This deprivation is caused by the 'globalisation and the lifestyle habits of the West' in general (p.4) and 'the repetitive and always-the-same taste of many industrial products, combined with the massive overuse of sweeteners, salt and artificial flavourings' in particular (pp.4–5). Overcoming sensory deprivation necessitates a new pedagogy for 'developing a balanced personality [...] able to orient themselves with greater awareness, and to use their trained senses to evaluate the consequences of our choices for their future' (p.5). 'Choice' is a key concept and the cartoon inserted as the conclusion helpfully specifies what is anticipated here. In this example, the personality who is able to consciously choose 'what to eat' and knows 'quality' is idealised as its educational goal. In addition, this quality is instantly explained and further concretised as 'Good' (healthy and organoleptically pleasing), 'Clean' (respecting nature and the environment) and 'Fair' (fair and ethical trade), all of which constitute the slogan of the Slow Food movement. In sum, the Slow Food method is defined as a pedagogy for developing good 'consumers' (p.4) who choose to eat not only for their own health, but also for their social environment, which includes all food system actors.

Similarly, the *Shinagawa* method is justified by pointing to children's 'impoverishment of the five tastes' due to the globalisation and delocalisation of food (Shinagawa 2013, p.9). Moreover, its nuanced difference from the French and Italian methods should be noted, namely that it repeatedly juxtaposes the 'five tastes' with mental qualities (e.g., 'Japanese mind', 'Eastern spirit', 'unique view of nature') as its educational objectives. To take a primary example of the latter, the concept of *hi-hu-mi-yo-i-mu-na-ya-ko-to* (fire-air-water-soil-rice-plant-insect-fish-bird-animal-human in the language of Old Shinto) is introduced to represent the Japanese cosmology, in which humans are dependent on all other non-humans and should thus be grateful for their own existence. It is anticipated that children will reach such a realisation through the awakening their five senses (pp.15–17). Although these nuanced differences between the three methods are, the important commonality of their justification and educational objectives can be identified as follows:

i *Justification*: Taste education was developed to address children's 'impoverishment of taste' due to the excessive medicalisation, industrialisation and globalisation of food.
ii *Educational objective*: The aim is to expand their abilities to control their own bodies and construct harmonious relationships with their social environments (including others).

3.2.2 Theories, Educational Content and Materials

The theoretical basis of the *Puisais* method is twofold. On the one hand, the physiology of taste is mobilised to explain how taste interacts with the other senses and how these signals are combined in the human brain. On the other hand, child psychology is also referenced to explain the development of their tastes and characteristic eating behaviours (such as dislikes and neophobia), in which the studies of Matty Chiva (1984), the pioneer of the French school of the psychology of food, comprises its important theoretical backbone. In the following chapter, the programme consisting of ten sessions (five senses, basic tastes, olfaction, vision, etc.) are described, each of which is generally structured according to the following schema: (1) exposition of educational content and underlying theories, (2) tasting food samples and (3) verbalisation of sensory experiences and communication with peers. Following this schema, the children are expected to learn the physiological mechanism of taste, as well as its inevitable relations with geographical, historical and cultural factors. Despite its total scope, the latter educational content has yet to be theoretically justified by any social theory in the textbook.

The immense similarity in the educational content can be identified with other two methods (see the corresponding numbers in Table 3.1). Still, there are also differences. The Slow Food method is also theoretically based on the physiology of taste, but not so evidently on child psychology (with only a partial psychological explanation of olfaction and some scholarly references [Logue & Freeman 1991], although their connection with the educational content is unclear). History (Capatti & Montanari 2003), anthropology (Harris 1998) and the sociology of food (Revel 1984; Fischler 1990; Lupton 1996) are cited in the bibliography, but, again, the relation of their theories to the educational content is hardly identifiable.

Compared to the high degree of similarity between the French and Italian methods, the *Shinagawa* method contains some unique educational and (quasi-)theoretical qualities. Instead of the child psychology, Japanese and Eastern thought is used to explain the cultural and spiritual aspects of sensory experiences. Aside from the aforementioned ecological illustration based on Old Shinto, the method also introduces the Confucianist food mentality ('*fuji-fushoku*', meaning not to eat that food if its right moment is missed) and the Buddhist sense of gratitude for food (expressed by '*itadakimasu*' [I humbly receive] and '*gochiso-sama*' [I thank the foods and the persons who prepared them], which are meant to be said before and after meals in Japan), as well as a particular emphasis on *umami* as the fifth basic taste. Nevertheless, such culturalist illustration remains occasional and awaits a more systematic theorisation. Again, these differences do not dilute the importance of the commonalities in their theories, educational content and materials, which can be summarised as follows:

- i *Theory*: The pedagogy is developed theoretically, based primarily on the physiology of taste and partially on child psychology. However, social theories are not used to justify the totality of the educational content.
- ii *Educational content*: The children learn primarily the physiological mechanism for perceiving taste (which is composed of the five senses), coupled with the relevant socio-cultural dimensions of food.
- iii *Educational material*: The children do not learn only from textbooks, but also engage in lively experiences of tasting and communicating with others.

3.2.3 *Teaching Methods and Principles*

The structure of the teaching method has already been discussed. Only one of its constituents, the 'verbalisation of one's sensory experience and its communication with others', is more closely analysed here. In articulating its unique teaching method, Puisais (1999) requires two principles. The first is that of *respect for subjective senses*, which not only reconfirms that 'each has his own taste' but also extends it to that of 'others' (p.44). The second is the principle of *relativisation*, which encourages the interpersonal comparison of senses to avoid any dogmatic subjectivity (p.39). Drawing on these two principles, the third principle of *non-normativity* is introduced in order not to presuppose any good or bad taste (p.44). These three principles are consistent with the educational objective of constructing 'a pleasurable relationship with one's own body and others'.

The Slow Food method also follows these three principles of communication by claiming that 'it is important to guarantee everyone the chance to freely express their own sensation without any preliminary judgment' (Barzanò & Fossi, p.7) and that 'collective evaluation is a fundamental step [...] to overcome the boundaries of subjective perceptions and tastes' (p.7). This is also the case in the *Shinagawa* method, which requires taste educators 'to encourage children's inquiry into their own senses and concepts without prescribing them with some predetermined concept of good senses' and 'to facilitate their communication' to assist such exploration (Shinagawa 2013, pp.106–109). In summary, the identified commonality can be expressed as follows:

i Teaching method: The pedagogy is comprised of theoretical exposition, tasting food samples and verbalisation or communication with others, the last of which follows the three principles of 'respect of subjectivity', 'non-normativity' and 'relativisation of own senses'.

3.3 Inherent Challenges as a Pedagogy for Developing 'Gastronomes'

3.3.1 *Differences in 'Good Subjects' and Their Underlying 'Gastronomic' Cultures*

I have so far discussed the commonalities of various taste pedagogies with a view to articulating their pedagogical identities and producing practical pathways for further institutionalisation. Nevertheless, the nuanced differences when normalising 'good subjects' in each pedagogy should not be dismissed.

In the *Puisais* method, good subjects are deemed good eaters with abilities to construct pleasurable relations with their own senses and others, that is, 'future gastronomes' (pp.8–10). This concept is polysemic and often misleading; for example, it is viewed as too hedonistic and elitist. To avoid such educationally unfavourable connotations, the term 'gastronome' is substituted with neutral terms such as 'persons who know how to eat well' in the Japanese translated textbook

(Ishii et al. 2016). Gastronomy is a key concept for understanding good subjects in the Slow Food method, in which normalising the notion of 'consumers-citizens' who know what Good, Clean and Fair is embodies the core value in the movement of 'neo-gastronomy' (otherwise 'eco-gastronomy' or 'populist gastronomy') (Sassatelli & Davolio 2010). At first appearance, the *Shinagawa* method seems to go in a completely different direction by setting those with 'Japanese views on nature' or 'Eastern minds' as its educational objective; however, these qualities are indeed essential constituents of *washoku* (UNESCO 2010), and, as I shall discuss in Chapter 9, Japanese gastronomy. In other words, these nuanced differences in normalising good subjects originate from the cultural variations of their underlying 'gastronomies', which should be scrutinised in future studies to arrive at the fundamental philosophical assumptions of taste education.

3.3.2 *Incomplete Distinction from Nutrition Education*

Generally, taste education is distinguished from traditional nutrition education. This consideration was also applied to the three pedagogies in question (Puisais 1999, p.11; Shinagawa 2013, p.106). However, during my textbook analysis, I was unable to identify any 'objective' differences in the content from nutrition education: (i) As for the educational content, both pedagogies are based, in a similar manner, primarily on physiological theories (whether taste or nutrition). It is certain that taste education includes a wider range of content (geography, history and culture); however, this total scope has yet to be theoretically justified.

Moreover, the recent enlargement of educational content in nutrition education itself is also complicating such pedagogical distinctions. (ii) As for the educational materials and teaching methods, taste education is claimed to be centred on children's 'experience' of tasting and communicating with others, while nutrition education is viewed as a 'knowledge-driven' approach. Yet, strictly speaking, this does not mean that lively experiences are completely omitted from nutrition education. Indeed, there are some programmes which favour experiential learning (the use of food samples, hands-on farming experience, etc.), as seen in Chapter 2. The problem here is that 'experience' in taste education lacks a theoretical foundation. If one aims to promote taste education in Japan, it is also necessary to address criticisms of experience-based educational theories, particularly concerning how children's experiences are integrated into a systematic body of knowledge (Shibata 2010a, 2010b). Readers might already consider John Dewey's experiential empiricist education for this purpose, and I shall also attempt to integrate Dewey into the theoretical construction of taste education in particular and food education in general (Chapter 10).

3.3.3 *Logical Contradiction Concerning the Principle of Non-normativity*

In relation to such a distinction, a more complex problem in taste education concerns its principle of non-normativity, which was initially proposed as a result of reflection on the limits of 'normative' nutrition education premised on nutritionally defined good or bad food. However, the taste educators in question failed to recognise the 'logical contradiction' inherent in their teaching methods. As was identified in the textbook analysis, taste education was (i) developed by problematising the globalised, industrialised and medicalised food, and by integrating certain norms regarding good or bad food. However, (ii) its adherents simultaneously claim to respect the principle of non-normativity. Moreover, even if, in theory, educators limit their normative judgements in the educational process, certain food samples used for the educational purposes (organic, home-made, healthy, etc.) can serve as 'hidden curricula' to have unintended normalising effects on the children.

Having acknowledged this logical contradiction again, it is necessary to construct a theoretical basis for the principle of non-normativity. One possibility is already suggested in critical studies of taste education, which mobilise Foucauldian or Bourdieusian theories to point to its negative consequences (Hayes-Conroy 2009; Leer & Wistoft 2018). I discuss the significance of such criticisms, as well as their limits, later in this book.

3.4 Concluding Remarks

As an entry point into the diagnosis of taste education as an alternative approach, the focus of this chapter was its pedagogical nature, which was determined through textbook analysis. The following commonalities were identified among the original and variant pedagogies: (1) *Justification*: The pedagogies were developed to address children's 'impoverishment of taste' due to the excessive medicalisation, industrialisation and globalisation of food. (2) *Objective*: They are aimed at expanding children's abilities to control their own bodies and construct pleasurable relationships with their social environments. (3) *Content*: Children learn primarily the physiological mechanisms of perceiving taste (which is composed of the five senses), coupled with the relevant socio-cultural dimensions of food. (4) *Method*: These pedagogies are comprised of theoretical exposition, tasting food samples and verbalisation or communication with others, which follows the three principles of 'respect of subjectivity', 'non-normativity' and 'relativisation of own senses'. (5) *Theory*: These pedagogies were developed theoretically based on the physiology of taste and child psychology.

The identified nature of this pedagogy can serve as common ground for articulating its general educational impacts on children, as well as encouraging possible cooperation between food education actors. However, certain challenges were also elucidated concerning the articulation of the link between taste education and gastronomy, the justification of sensory experience in the educational process and the logical contradiction regarding the principle of non-normativity. All of these require an expansion of their theoretical basis, from the current bio-psychological to the social and cultural theories.

Notes

1 This chapter was reprinted with permission from Ueda, H. (2020) History and pedagogical nature of taste education in Japan, France and Italy through textbook analysis. *Journal of Food System Research*, 27(2), 48–66.

2 Although Venturi and Peris developed the unique method, this 'uniqueness' did not come from nowhere. According to these two, whom I interviewed in September 2016, one cannot deny the influence of the prestigious Italian pedagogy, Montessori education, which also puts sensory experience (not limited to tasting) at the pedagogical centre.

References

Barzanò, C. & Fossi, M. (2010) *In What Sense? A Short Guide to Sensory Education*. Slow Food International.

Capatti, A. & Montanari, M. (2003) *Italian Cuisine: A Cultural History*. Columbia University Press.

Centro Educazione del Gusto & Nakano, M. (2012) *Mikaku no Gakko [School of Taste]*. Kirakusha.

Chiva, M. (1984) *Le Doux et l'Amer*. Presses Universitaires de France.

Fischler, C. (1990) *L'Homnivore*. Odile Jacob.

Fujioka, N. (1991) *Shakai Ninshiki Kyoikuron [Educational Theory of Social Recognition]*. Nihon Shoseki.

Harris, M. (1998) *Good to Eat: Riddles of Food and Culture*. Waveland Press.

Hayabuchi, H., Ueda, H., Umeki, Y., Eto, W. & Ota, M. (2017) Effects of the 'Lesson of Taste' on taste recognition, food awareness, and eating behaviors among elementary schoolchildren. *Journal of Japanese Society of Shokuiku*, 11(4), 323–333.

Hayes-Conroy, J. (2009) *Visceral Reactions: Alternative Food and Social Difference in Two North American Schools*. Pennsylvania State University, Ph.D. Thesis.

Ishii, K., Puisais, J., Sakai, N. & Tajiri, I. (2016) *Kodomo no tameno Mikaku Kyoiku [Taste Education for Children]*. Kodansha.

Ishii, K. & Tajiri, I. (2017) *Kodomo no Mikaku wo Sodateru: Oyako de Manabu Puisais Riron [Developing Children's Taste: Learning the Puisais Theory in Family]*. CCC Media House.

Leer, J. & Wistoft, K. (2018) Taste in food education: A critical review essay. *Food and Foodways*, 26(1), 1–21.

Logue, A.W. & Freeman, W.H. (1991) *The Psychology of Eating and Drinking*. Freeman.

Lupton, D. (1996) *Food, the Body and the Self*. Sage Publications.

Miyagawa, J. (2016) *MIIKU Master Course Textbook (4th ed.)*. MIIKU.

Montoux, D., Martin-Poly, C. & Reverdy, C. (2012) *Programme de Formation des Enseignants sur les Classes du Goût*. Ministère de l'Agriculture de l'Agroalimentaire et de la Forêt & Ministère de l'Education Nationale, de la Jeunesse et de la Vie Associative.

Puisais, J. (1999) *Le Goût chez l'Enfant : l'Apprentissage en Famille*. Flammarion.

Puisais, J., MacLeod, P. & Politzer, N. (2002) *Le Goût et les 5 Sens*. Cédérom Odile Jacob Multimédia-Screen.

Revel, J.F. (1984) *Culture and Cuisine: A Journey through the History of Food*. Da Capo.

Sasaki, T. (2008) *Slow Food Suginami Tokyo Sengen [The Declaration by Slow Food Suginami Tokyo]*. Siryusya.

Sassatelli, R. & Davolio, F. (2010) Consumption, pleasure and politics: Slow Food and the politico-aesthetic problematization of food. *Journal of Consumer Culture*, 10(2), 202–232.

Sato, M. (2009) Mikaku kyoiku wo toriireta chori gino shutoku no jyugyo dukuri [Development of the cooking class integrating Classe de Goût]. *Journal of Japan Association of Home Economics Education*, 51(4), 310–314.

Schlich, P., Lange, C., Gaignaire, A., Reverdy, C. & Politzer, N. (2014) *Effet d'une éducation sensorielle sur les préférences et les comportements alimentaires d'enfants âgés de 8 à 10 ans*. Eveil'O'Goût.

Shibata, Y. (2010a) *Gendai no Kyoikugaku [The Contemporary Educational Theories]*. Gakubunsha.

——— (2010b) *Kyoka Kyoikuron [The Theories of Subject Education]*. Gakubunsha.

Shinagawa, A. (2013) *Food Consciousness Textbook for Ajiwai Method*. Food Consciousness Institute.

Togawa, R. (2009) France no shogakko ni okeru shokuiku [Food education in French elementary schools]. *Benesse Educational Research & Development Center*, 15, 40–45.

Uchisaka, Y. (2007) *Kodomo no Gokan wo Mezamesaseru Mikaku no Jyugyo [Leçon de Goût Awakens Children's Five Senses]*. Godo Shuppan.

Ueda, H. (2017) Impact evaluation of a school-based food education intervention on cognitive and attitudinal aspects of children's dietary habits: A case study of a sensory education program 'Leçon de Goût'. *Journal of Food System Research*, 24(1), 2–17.

——— (2020) History and pedagogical nature of taste education in Japan, France and Italy through textbook analysis. *Journal of Food System Research*, 27(2), 48–66.

United Nations Educational, Scientific and Cultural Organization (2013) Washoku; traditional dietary cultures of the Japanese, notably for the celebration of New Year (nomination file, No. 00869).

4 Educational Effects on Children

Initiated in the medical sciences in the 1990s (Guyatt 1991), the 'evidence-based' approach has also been utilised for food education. However, there is still a gap between the ideal and the actuality. A comprehensive review of food education in the medical sciences (Tsuji et al. 2011) revealed the somewhat shocking fact that educational effects had not been evaluated in most programmes. The interdisciplinary review in Chapter 2 identified a gradual improvement of the situation (notably in nutritional sciences), but that many programmes still suffered from low-evidence approaches (ad-hoc choice of indicators, the absence of pre-post evaluation, etc.). Moreover, there was no study in which the evidence-based approach was employed in the true sense, whereby food education programmes should continuously evolve through a cycle of practice, evaluation, improvement and re-evaluation. Taking taste education as a primary example, in this chapter an ideal type of evidence-based food education is demonstrated.[1]

Of the various pedagogies identified in the previous chapter, the Japanese LSG method was selected as a focus of study. The LSG method is a shortened version of the original *Puisais* method into one or two sessions and is practised mainly within the framework of the annual national *La Semaine du Goût* (The Week of Taste). Although it would be ideal to examine the *Puisais* method from an educational perspective, the focus here is to demonstrate how to construct the evaluative system. Thus, the LSG method was favoured from a practical perspective. Its growing popularity in Japan is another advantage of this choice (from 58 classes in 2011 to 591 classes in 2019).

4.1 Development of Evaluation Systems

To establish the evaluation system for the Japanese LSG method (also called *Leçon de Goût*), two sets of literature were mobilised and all the potential educational impacts were identified. The first item was the guideline developed for taste educators, which stipulated its basic principles (Kubota et al. 2011). The second was relevant research articles in which the educational impacts of taste education on children in different settings were explored. These expected educational impacts are summarised in Table 4.1, followed by further selection and modification. Note that there was no reliable method for 'pleasure of eating' and its evaluation was not performed this time.

DOI: 10.4324/9781003341895-6

Table 4.1 Summary of Expected Educational Impacts

Educational Impacts	*Guideline*	*Research Articles*	
Taste mechanism	○	○	Improved (Tokumaru et al. 2014)
Pleasure of eating	○	–	
Food differentiation	○	○	Improved (Mustonen et al. 2009; Hayabuchi 2015)
Taste expression	○	○	Improved (Mustonen et al. 2009; Reverdy et al. 2010; Tokumaru et al. 2014)
Taste preference	–	○	Increased preference for complex flavours (Reverdy et al. 2004)
Food neophobia	–	○	Temporally decreased (Reverdy et al. 2008), Decreased (Park & Cho 2016)
Food curiosity	○	○	Increased but not significant (Gaignaire et al. 2011)

4.1.1 *Understanding the Taste Mechanism*

The basic four or five tastes (sweet, sour, bitter, salty and *umami*) account for only a part of 'taste'. Rather, a sense of 'taste' is created through the interaction of the five senses (taste, vision, hearing, smell and texture) and even emotions, as well as cultural and physiological conditions. Thus, a basic understanding of this complicated taste mechanism is essential to effectively conduct tasting practice and acquire other educational impacts. The measurement method of Tokumaru et al. (2014) was employed, whereby children were asked open-ended questions about the five senses and five basic tastes, and the number of correct answers about each taste and sense was counted.

4.1.2 *Food Differentiation*

The ability to differentiate food ingredients and flavours is one of the indicators of taste sensibility. Mustonen et al. (2009) used multiple sensory evaluation techniques in laboratory settings, such as free odour naming and taste identification of six solutions, which is, however, not suitable for the practical and feasible evaluation system in normal school settings. Instead, the dietary record-based model of Hayabuchi (2015) was applied in this study, whereby the children were asked to identify the names of foods and the seasonings used in the school meals. To minimise the effect resulting from the variety of school meals (12–19 ingredients), the average score was calculated from the correctly identified foods for three consecutive days before and after the programme.

4.1.3 *Taste Expression*

To quote Puisais, there is no taste if one has no word, and vice versa. To evaluate the ability to express taste, the method developed by Gaignaire et al. (2011) was employed, in which children were asked to describe the tastes and sensory impressions of school meals on the dietary recording sheets for six days during the program. Their description texts were then analysed and compared, with a special focus on the five basic tastes and other taste-related nouns, adjectives, adverbs, *na*-adjectives (*keiyo-doshi* in Japanese) and the types and frequency of these words.

4.1.4 *Taste Preference*

The basic premise of taste education is that a taste preference for more complexity and less intensity enables children to lessen their excessive desire for sugar and fat, and discover a greater variety of flavours (Reverdy 2011). Following this principle, children's declarative (not behavioural) preferences for taste complexity and intensity were used as indicators to assess how the programme managed to shift their taste preferences. The children were asked to indicate the degree of complexity and intensity they considered the most favourable for them on a scale of 1–4 ('very simple/weak' to 'very complex/intense'). In addition, open-ended questions about their associations with pleasant and unpleasant tastes were asked to holistically comprehend their taste predilections.

4.1.5 *Food Neophobia*

Food neophobia is a personality characteristic and behaviour involving the rejection of novel food (Pilner & Salvy 1971). It can be measured by asking children about their willingness to taste novel foods or anxiety about doing so, such as with a questionnaire using an adapted food neophobia scale (AFNS) (Reverdy et al. 2008). A version of the AFNS translated into Japanese was used for this study (see Appendix for the questions items).

4.1.6 *Food Curiosity*

Compared to food neophobia, curiosity about food (or food curiosity) is not an established concept, despite its intrinsic importance in taste education. Gaignaire et al. (2011) operationally defined it as 'the ability of the eater to want to know everything that is related to food, whether at the stage of production, processing or consumption'. Based on this definition and on the experience of conducting the EduSens survey (2005–2009), a measurement scheme comprising eight attributes (extending from shopping to cooking and eating) of food curiosity was developed. The translated version in Japanese was also utilised for this study (see Appendix).

4.1.7 *Evaluation Design*

The survey for the impact evaluation was carried out with 97 children (aged 10–11 years) at an elementary school in Kyoto in October 2015. The pre-post evaluation was designed using a questionnaire (for taste mechanism, taste preference, food neophobia and food curiosity) and a dietary record survey on school meals (for food differentiation and taste expression). Children who had food allergies or could not be present for the entire evaluation process were excluded from the analysis. A total of 81 students (83.5%) completed the questionnaire, while 63 students (64.9%) filled in the dietary records. Each score obtained in the evaluation before and after the programme was statistically analysed utilising SPSS (Version 20.0, level of significance: $p < 0.05$).

The programme of *Leçon de Goût* (Figure 4.1) consisted of the general educational content and teaching methods, as identified in the previous chapter: (1) In the theory-learning part (45 mins), the children first studied the taste mechanism composed of the five senses. They then moved on to tasting basic tastes, utilising food samples such as sugar, salt, rice vinegar and bitter chocolate. (2) In the applied tasting part (45 mins), the children discovered the new tastes and flavours of various chef-made products (roast pork with apple pie, and matcha-flavoured chocolate cookies), and (3) they communicated these sensory experiences to their peers and the chef-teacher.

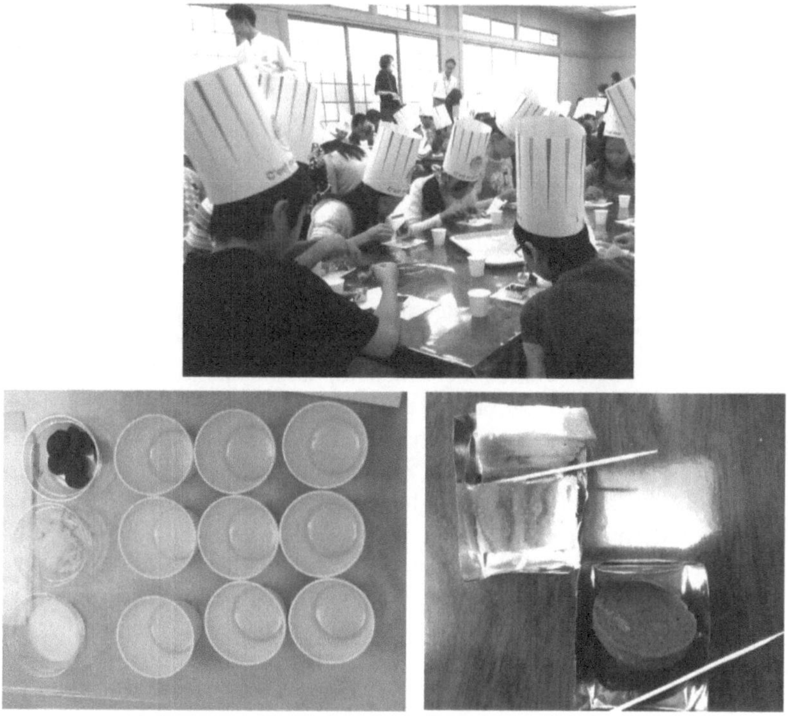

Figure 4.1 Leçon de Goût Performed in an Elementary School in Kyoto

Source: Photographed by the author during the intervention

Note: Children tasting two types of food samples, one of which has four basic tastes (sugar, salt, rice vinegar and bitter chocolate) and another comprising samples (matcha-flavoured chocolate and roast pork with apple pie) made by the chef, while wearing the special *toque blanche* (hat for chefs).

4.2 Evaluation of Educational Impacts

4.2.1 *Improved Abilities and Limitations*

The educational effects were observed for their understanding of the taste mechanism and, partially, for taste expression, taste preference and food curiosity. I will discuss taste expression in more depth below. Here, I focus on the three other effects.

Taste mechanism: A statistically significant increase was observed in the level of understanding of the taste mechanism (five senses: 1.2 to 3.2 out of 5 points, basic tastes: 0.63 to 3.11 out of 4 points).[2] The lack of knowledge of taste prior to the programme was also demonstrated, as well as the ability of taste education to fill this gap in the existing school-based food education.

Taste preference: Although no statistically significant change was observed for their preference for complex (p = 0.18) and light flavours (p = 0.18), further analysis of their open-ended answers demonstrated an increase in their preferences for lighter tastes, as well as a decrease in their preferences for sourness. Interestingly, the former is the intended effect of taste education, while the latter is the opposite of its assumption. This contradiction might have been due to the materials used for the programme, such as presenting only (industrial) rice vinegar as an example of sour food for guided tasting. There would have been different consequences if more educationally suitable foods (such as fresh orange) were used. This case indicates the importance of carefully choosing the educational materials (and judiciously considering their normativity).

Food curiosity: There was a statistically significant increase in the score for curiosity about food (2.86 to 3.02). However, it must be noted that the internal consistency of the question items used was 0.6 at best, while an internal consistency of 0.7 or higher is usually considered 'reliable'.

On the other hand, no statistical significance was observed for food neophobia (p = 0.18) and food differentiation (p = 0.21). For the former, the score increased from 4.17 to 4.30 after the programme, which implies a reduction of food neophobia. This relatively weak educational impact might have been due to the choice of food samples (roast pork and matcha-flavoured chocolate), which were too familiar to the children and, therefore, not particularly suitable for challenging their fear of novel foods. For the food differentiation, interestingly, there seems to have been a drop in the correct answers (40.8% pre-programme and 32.5% post-programme). The latter was inevitable because the dietary record on every school meal was too demanding for the children, who had difficulty spending enough time and concentrating on the recording while curbing their enthusiasm for break-time playground activities after the meal.

4.2.2 *Lexical Analysis of Taste Expression*

The reason for which I separated out the discussion on taste expression was that I wished to present here a possibility of employing a more elaborate form of analysis than that mentioned above: (i) As initially planned, the basic tastes used in the dietary records were analysed. There were no statistically significant changes in the children's word choices. As is the case with food differentiation, this was probably due to their distraction during the evaluation. However, if the focus is extended to a wider range of taste-related words, some nuanced educational impacts can be observed. (ii) Further analysis was made of the subjective expression '*oishii*' (good), the frequency of which significantly increased from 46% to 61% after the programme, in contrast to the relative decrease in other objective word choices (Table 4.2).

This finding is consistent with a previous study conducted in France (Gaignaire et al. 2011), in which it was found that taste expression tended to be more subjective in relaxed contexts (such as at the family table). Nevertheless, the educational effect of taste education should not be denied, for there is still a possibility that the nature of this subjective expression would have been transformed. To investigate this, co-occurrence network analysis was conducted on the expression '*oishii*', using a lexical analysis software tool (called KH Coder). As illustrated in Figure 4.2, the network around '*oishii*' expanded after the programme

Table 4.2 Taste Related Vocabularies Used during School Meals

	Pre-programme	*Type*	*Freq.*	*Post-programme*	*Type*	*Freq.*
Nouns	Sweetness (n = 4) Flavour (3), Aroma (2), *Umami* (2) Bitterness, Sourness, Depth (1)	7	14	Texture (n=3) Sweetness (2)	2	5
Adjectives	**Good (*oishii*) (n = 150 in a total of 326)** Good (*ii*) (39), Sweet (21), Soft (21) Intense (12), Pungent (5), Weak (4) Sour (3), Hard (2) Bitter, Salty, Small, Bad, Fragile (1)	14	262	**Good (*oishii*) (n = 155 in 254)** Good (*ii*) (23), Sweet (7) Intense (7), Weak (2), Gentle (2) Warm, Sweet & Sour, Pungent Sour, Bitter, Oily, Many, Too Much	14	204
Na-Adjectives	Smooth) (n = 2) Hot, Creamy, Juicy (1)	4	5	Mushy, Creamy, Pungent (1)	3	3
Adverbs	Crispy (n = 20), Heavy (10) Light (6), Spongy (2) Crunchy, Dried, Fresh, Pop, Thick, Fluffy, Freshly-baked (1)	11	45	Crispy (*shaki-shaki*) (n = 19), Heavy (*shikkari*) (5), Light (5) Freshly-baked (2), Crunchy (2) Crispy (*pari-pari*) (2), Thick (2) Dried, Fresh, Light & Crispy Heavy (*kotteri*), Simple (1)	12	42

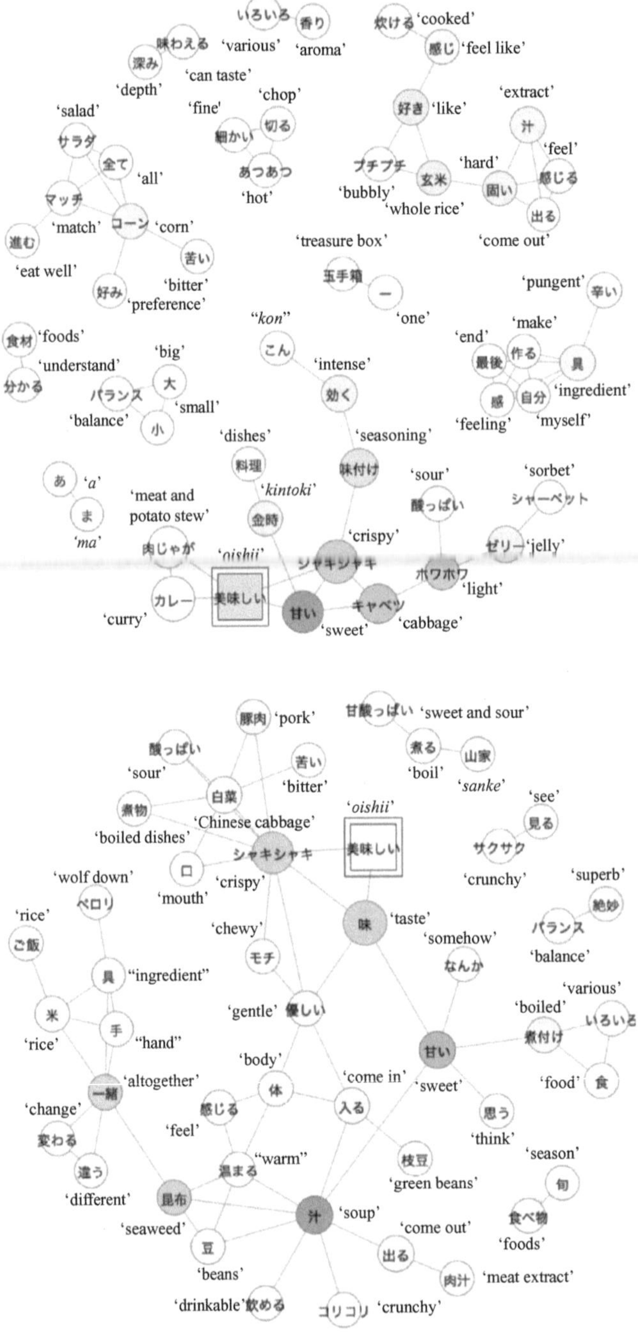

Figure 4.2 Co-occurrence Network around the Word 'Oishii'

Source: Author's analysis using KH coder. Reprinted with permission from Ueda (2017, p.12).

Note: The upper figure shows the co-occurrence network around the word *oishii* (in the square box) before the programme, whereas the bottom shows the network after the programme. Only words with a frequency of more than three utterances are included here. The words with higher frequency are represented by thicker circles.

(14–35 nodes), suggesting that the programme enabled the children to use more varied vocabulary in combination with '*oishii*'. In addition, concordance analysis was performed to elucidate how '*oishii*' had been actually used in taste expression phrases, revealing the five semantic categories in which the children used the word '*oishii*': (1) texture (e.g., 'it's *oishii* because the meat is soft'); (2) combination (e.g., 'it's *oishii* because the meat juice and rice match very well'); (3) taste intensity (e.g., 'its *oishii* because the sesame dressing on vegetables tastes light'); (4) the nature of the taste and food (e.g., 'the curry flavour of meat and potato stew is *oishii*'); and (5) variation of taste (e.g., 'it's *oishii* because the taste changed when eating them together'). In particular, a statistically significant increase of frequency was observed for the texture category after the programme. In sum, the programme made it easier for them to perceive, whether consciously or unconsciously, something *oishii* based on its texture (crispiness or softness). In this sense, the children's taste expression abilities were enhanced to some extent.

4.3 Improvement of the Educational Programme

4.3.1 *Programme Duration and Coupling with School Meals*

To embody the evidence-based approach, food education programmes should continuously evolve through a cycle of practice, evaluation, improvement and re-evaluation. In the third phase of this study, three types of improvement were made drawing on the abovementioned evaluation. The first update related to their limited duration. To overcome this, the *Leçon de Goût* programme, which originally consisted of two sessions, was coupled with school lunchtimes (a total five days). In Japan, lunch break is 45 minutes, during which mealtime generally takes 15–20 minutes, excluding preparation before the meal, and cleaning and playground activities after the meal. Although it is not so ideal to perform a complete lesson within this time range, it is consistent with the recent revaluation of school lunchtime as an educational opportunity (Chapter 1) and it can also overcome the low feasibility of introducing long sessions into the pre-determined educational curricula (as pursued in Chapter 5).

4.3.2 *Taste Sensibility and the 'Meaning' of Sensory Experience*

Three improvements were made in terms of the evaluation system, two of which were relatively minor, while the other required careful consideration. First, the measurement of food differentiation as an indicator of taste sensibility was replaced with sensory analysis because, on the one hand, the children had struggled to complete effectively the task related to the former in the previous study. On the other hand, a similar study on the original *Leçon de Goût* programme (Hayabuchi et al. 2017) recently determined the effect on children's taste sensibility using the latter method. In the sensory analysis, a valid method often used for professionals, the children were presented, under their teachers' oversight, with the test solutions of sucrose 0.4% (sweet), citric acid 0.02% (sour), sodium chloride 0.13%

(salt), anhydrous caffeine 0.025% (bitter) and sodium glutamate 0.05% (*umami*) before and after the intervention.

The second modification was necessitated due to cultural differences related to the act of tasting. In contrast to the French gastronomic culture, in which tasting and its communication is pursued, the Japanese food culture has historically not favoured verbal communication during the meal, including that of sensory experience, mostly due to Buddhist and Confucianist influences. The effects of such religious principles and mores are not consciously perceived in contemporary Japan. Nevertheless, most children in the previous study faced some confusion and hesitation about communicating taste experiences with their peers. Therefore, at this introductory stage, it was fundamentally important to investigate what 'meanings' Japanese children ascribed to the act of tasting, rather than rushing to short-term results. In terms of this logic, the measurement of taste preferences, which should take a considerable time to formulate, was replaced with that of representation of tasting: In the post-programme questionnaire survey, the children were asked the open-ended question 'what does tasting mean to you?' and their answers were analysed in relation to the educational content.

4.3.3 Development of Food Curiosity Scale

Third, food curiosity in the previous study was measured based on the tentative definition and method of the French study (Gaignaire et al. 2011), but its internal consistency did not satisfy an acceptable level for practical use (Cronbach α: 0.61). Therefore, a revised food curiosity scale was developed from a psychological perspective, which generally involves the following steps.

1 *Theoretical Justification*: The literature on educational psychology was reviewed to formulate a definition of food curiosity. According to Loewenstein's (1994) wide-ranging review of theoretical perspectives on 'curiosity', the earliest thinkers saw it as an 'intrinsically motivated desire for information' (Aristotle), a 'passion for learning' (Cicero) and 'thirst for knowledge' (Freud), as if it were a unidimensional concept. It was not until the 1950s that its multi-dimensionality started to be recognised. Berlyne (1954) pioneeringly located curiosity in two dimensions: one ranging between perceptual and epistemic curiosity, and the other spanning specific and diversive curiosity. Perceptual curiosity refers to a drive that is aroused by novel stimuli and reduced by continued exposure to these stimuli, whereas epistemic curiosity refers to a desire for knowledge and is applied mainly to humans. Specific curiosity refers to the desire for a particular piece of information, while diversive curiosity refers to a more general seeking of stimulation that is closely related to boredom. Other than Berlyne's fourfold categorisation of curiosity, the most commonly studied division of curiosity has been that between state and trait curiosity. State curiosity refers to curiosity in a particular situation, whereas trait curiosity refers to a general capacity or propensity to experience curiosity. As Loewenstein continues, although individuals with high trait

curiosity probably make superior students and scientists, the ability to measure such differences would, at best, aid in sorting or tracking students or scientists on the basis of their curiosity, whereas an improved understanding of state curiosity has the potential to suggest practical educational methods of stimulating curiosity in the broader population. In light of this, it is plausible to evaluate the development of an 'epistemic, specific and state curiosity' in our educational intervention.[3] Thus, in this study, an operational definition of food curiosity is tentatively proposed as 'a desire for a food-specific piece of information in a particular situation'.

2 *Selection of Question Items*: The next step was to develop an inventory of question items to measure food curiosity, with careful attention being paid to the style of the description and question specificity. With respect to the former, a number of existing curiosity scales (Leherissey 1971; Boyle 1979; Naylor 1981) were referred to, so there was no possibility of confusion with trait curiosity in the interpretation, particularly with reference to the children's state 'at a particular moment in time'. As for the latter, careful attention was paid to avoiding a unidimensional, biased interpretation of 'food'. Food-specific items were divided into the three domains employed by Gaignaire et al. (2011), namely production (including purchase), processing (including cooking and preparation) and consumption. Then, the following materials were referenced to concretise the items of each dimension: (i) the existing items in the previous study (Gaignaire et al. 2011), (ii) Japan's Basic Law on Food Education of 2005 and (iii) preliminary interviews with the school teachers who participated in this study. A total of 26 questions were tentatively selected, comprising 11 items in the production domain, 5 in preparation and 10 in consumption.

3 *Validation and Elaboration*: Before the pre-programme evaluation, the questionnaire survey was conducted among children (aged 9–12, n=125, of which 123 students [98.4% of the total sample] completed all the questions). Through factor analysis of their answers, the initial 26 questions were screened and reduced to 12 items, and three constructs of food curiosity were identified (Figure 4.3): The *Culinary Ability Acquiring* construct consisted of four items, which referred to a desire to acquire culinary skills and knowledge. The *Food Knowledge Seeking* construct consisted of five items, which more generally referred to a desire for knowledge concerning the bio-socio-cultural aspects of food (such as tastes, food cultures and food systems). The *Novelty Seeking* construct, consisting of three items, represented a desire to explore unknown or novel foods and tastes. In this way, the revised food curiosity scale, comprising 12 question items (Appendix), was finally developed (see Ueda 2018 for further details of this statistical validation).

4.3.4 Revised Educational Content

Together with these changes in the programme duration and evaluation system, four modifications or additions were also made to the educational content:

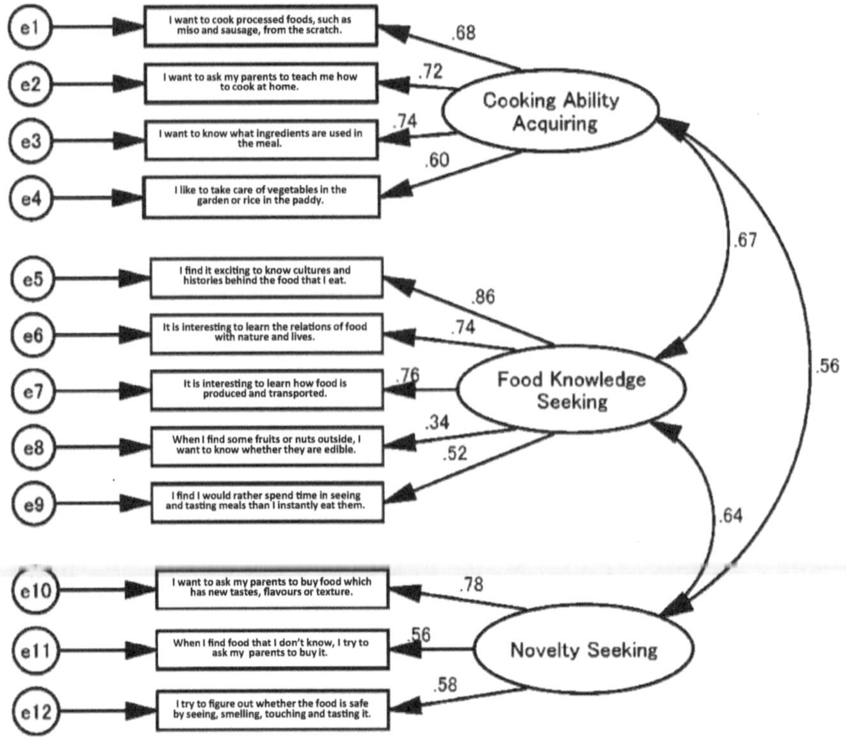

Figure 4.3 Food Curiosity Scale, Its Question Items and Three Constructs

Source: Structural equation modelling by the author using SPSS Amos (Version 22.0). Reprinted with permission from Ueda (2018, p.118).

Note: A confirmatory factor analysis demonstrated relatively high goodness-of-fit for this model (Goodness of Fit Index: 0.925, Comparative Fit Index: 0.971, Room Mean Square Error of Approximation: 0.046), while the internal consistency of these items was also proven acceptable enough (Cronbach α: 0.844 [for the entire scale], 0.789 [for Culinary Ability Construct], 0.768 [for Food Knowledge Construct], and 0.663 [for Novelty Construct]). E: error variable (n=123).

1 *Taste Mechanism*: Having proved that the existing programme could improve children's understanding of the taste mechanism, its physiological explanations (originally concerning 'how taste is perceived') were also extended to 'what physiological meaning each taste carries'.[4] The content also covered the fifth taste, *umami*, by using *katsuo-bushi* (dried bonito) as characteristic of Japanese food culture.
2 *Taste Expression*: In the existing programme, the children were encouraged to describe their sensory experiences, but not provided with any concrete strategy to extend their vocabularies. Therefore, a more systematic approach was adopted in the revised programme, in which the children were provided with useful strategies to construct the taste expressions stipulated in the professional textbook (Seto et al. 2005), such as advanced descriptions of the

five tastes (e.g., a roasting aroma through the nose) and of the degree and combination of the five basic tastes (e.g., moderately sour-sweet), as well as with information on the benefits of such taste expression, both in terms of better communication and keener sensibility.

3 *Food Neophobia*: In response to the relatively ordinary food samples used for the initial programme and its resultant weak effects, this revised programme included a more exotic food sample – an edible insect – to reinforce the educational impact on food neophobia. This was no doubt unfamiliar to most children; thus, to familiarise them with insect-eating before tasting, a modern culinary paradigm favouring edible insects (such as NOMA in Copenhagen) and its ecological benefits (Huis et al. 2013) were introduced with relevant documents and video materials. As an educational food sample, *inago-no-kanroni* (sweetened boiled locusts), a traditional delicacy unique to some rural regions of Japan (which also used to be consumed during the wartime food shortage), was prepared for this programme.

4 *Artisanal Products*: To avoid any possible adverse effects from the non-educational food samples, the tasting materials were mostly replaced with artisanal products (including traditional raw sugar, natural sea salt, mild-flavoured rice vinegar, freshly-shaven dried bonito and cacao powder without sugar). In addition, blind-tasting tasks were incorporated into the programme, using both artisanal and industrial soy sauces, with complementary information about their differing manufacturing processes, background histories and cultures, in order to encourage the children to extract some value from such traditional food cultures (Figure 4.4).

4.4 Re-evaluation

In terms of the evaluation design, three differences from the first study should be noted: First, in pursuit of more solid evidence, the programme using the non-randomised design was carried out with 43 children (aged 10–11) at an elementary school in Osaka in May 2017. Second, as the cost of this evaluation design, the size of each group had to be relatively small, comprising 20 students in the intervention group (90.9% of the total 22 samples) and 19 in the control group (90.5% of 21 samples), who completed the questionnaire surveys, sensory analysis and dietary records. Third, as discussed above, two previous indicators (food differentiation and taste preference) had to be replaced with two new ones (taste sensibility and representation of tasting), as well as the revision of the food curiosity scale. Because of such differences, the purpose here was not to compare the effects of the initial and revised programmes, but to verify whether the latter programme worked well.

4.4.1 Improved Abilities and Limitations

The educational effects were observed with respect to their food curiosity, understanding of taste mechanism, taste sensibility and taste expression, although each measurement required a careful interpretation. *Food curiosity*: An increase

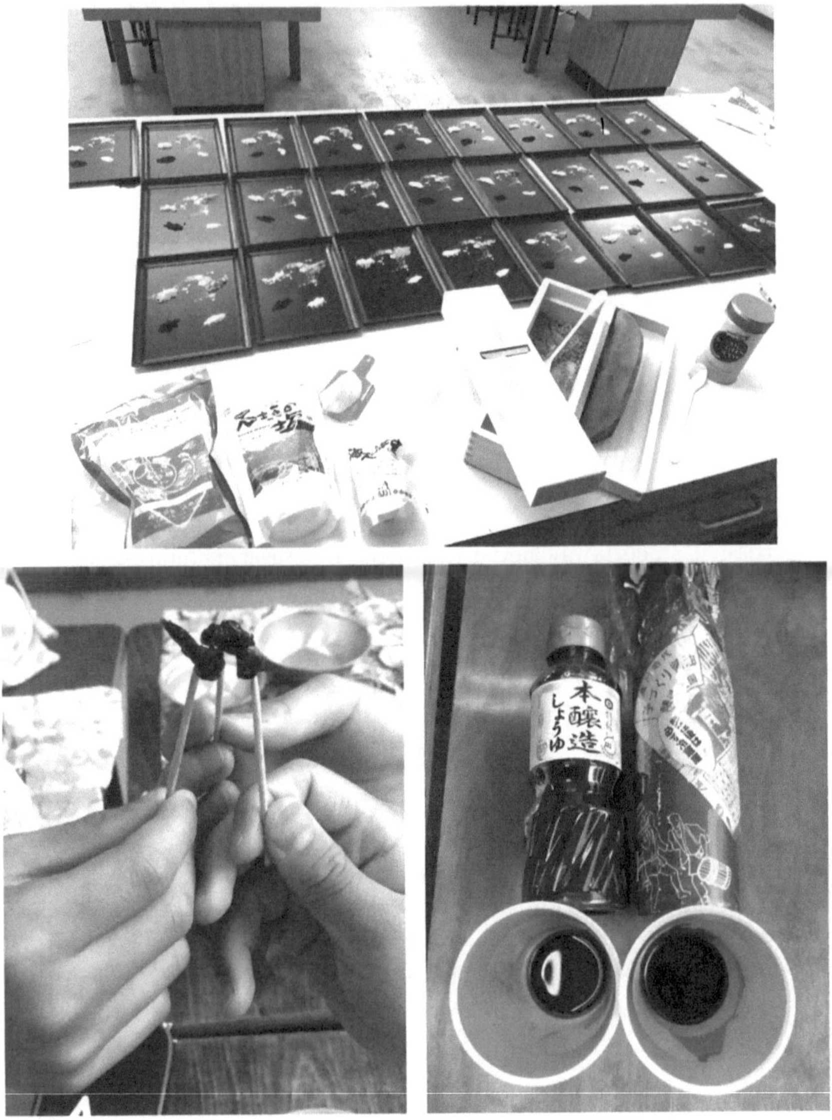

Figure 4.4 Revised *Leçon de Goût* Programme Performed in an Elementary School in Osaka

Source: Photographed by the author during the intervention

Note: The tasting samples, 'cheers with insects' and industrial (left) and artisanal (right) soy sauce.

was observed for the scale as a whole ($p < 0.05$), but not for each of the three constructs, which complicated the attribution of its effect to particular educational content. Yet, one can assume that, for example, the *Novelty Seeking* curiosity was stimulated by the experience of insect-eating. *Taste Mechanism*: Interestingly, there was an increase not only in the scores of the intervention group, but also in those of the control group, although the former's scores were higher after the programme (five senses and five tastes: +3.5 and +3.1 for intervention, +1.3 and +2.3 for control). Thus, positive effects were implied. Future studies require a larger sample to verify this finding.

Taste Sensibility: Similarly, increases were observed in the integrated scales, both for the intervention and control groups, so additional analysis was performed on each taste. There was an increase in three tastes (sweet, *umami*, bitter) in the intervention group and two (sweet, *umami*) in the control group, which proved its effect on their taste sensibility, at least towards bitterness. *Taste Expression*: (i) Of the five basic tastes, the intervention group experienced increases in frequencies of '*umami*' ($p < 0.05$), as well as of 'sweet', 'salty' and 'sour', although the latter three were not statistically significant.[5] (ii) As in the first study, a holistic analysis was conducted of all the taste-related words (nouns, adjectives, adverbs and *na*-adjectives), showing an interesting effect, namely that, despite their difficulty concentrating on one-week-long dietary records, the intervention group manifested a decrease in the use of subjective words (such as '*oishii*': 88 to 44 frequencies in the post-programme phase), while frequencies of objective expressions remained unchanged. This not only supported the premise of taste education but also demonstrated the success of revised educational content (i.e., a systematic approach to expanding their vocabularies).

On the other hand, the food neophobia scale did not increase in this study, although there seemed to be a difference between the intervention (+0.19 points) and control groups (+0.07) in the AFNS' absolute values. This result was surprising, given that the edible insects seemed to challenge effectively the children's neophobia at the beginning, but they later overcame it. This was evident in some of their enthusiastic reactions (e.g., saying, 'I ATE INSECTS' and 'honestly, that's more yummy than I expected'). Aside from the sample size, it might be advisable to reflect on the use of an alternative scale (tentatively developed under the Japanese condition by Imada and Yoneyama [1998]) to the AFNS, which had been initially developed in a Western context.

4.4.2 Representation of Tasting among Japanese Children

In answering the open-ended question 'what does tasting mean to you?', four major semantic categories were identified with respect to the children's representation of the act of tasting: (i) 'gratitude' towards living animals and the people who served the meals (Intervention: 22%, Control: 47%), (ii) 'pleasure' of eating (I: 11%, C: 17%), (iii) 'survival' such as satisfying appetite (I: 26%, C: 11%) and (iv) 'communication' of one's views to others (I: 21%, C: 6%). Interestingly, while the latter three categories corresponded with the general assumptions of taste education (i.e., the importance of physiological necessity, pleasure and socialisation, as indicated in Chapter 3), the first but most frequent one, expressing 'gratitude' to humans and non-humans through tasting, was somewhat unique and not effectively addressed in the French/Italian methods (including the revised programme here). To recall the previous chapter, only the *Shinagawa* method in Japan carefully focuses on this spiritual aspect, but without a theoretical justification. This challenge invites a more thorough analysis of the cultural tenets of taste education, a part of which is discussed in Chapter 9 in relation to the Japanese view of nature.

4.5 Concluding Remarks

In contrast to the increasing call for 'evidence-based' food education, there was still a gap between ideal and actuality. Taking taste education as a primary example, in this chapter, I demonstrated an ideal type of evidence-based food education comprised of a cycle of practice, evaluation, improvement and re-evaluation. First, based on a comprehensive literature review, a new evaluation system was developed to determine its educational impacts, including understanding taste mechanisms, food differentiation, taste expression, taste preference, food neophobia and food curiosity. Second, the intervention (the programme called *Leçon de Goût*) was conducted with children aged 10–11 in Japan (n=97) and four of the intended impacts were identified: understanding taste mechanisms, taste expression, taste preference and food curiosity. Third, the intervention programme was revised by drawing on the first evaluation, particularly concerning (i) the extension of the programme and coupling it with school lunchtime; (ii) the elaboration of the 'food curiosity' scale based on theories of educational psychology; and (iii) the refinement of the educational content. Fourth, the upgraded programme was conducted with children aged 10–12 (intervention: n=22, control: n=21) and the validity of four of the intended impacts were proved: understanding of the taste mechanisms, taste sensibility, taste expression and food curiosity.

This cycle must go on, but the case described in this chapter constitutes a rare demonstration of a consistent embodiment of evidence-based food education. This attempt also revealed further challenges, particularly concerning 'how to justify the programme's evolution'. For example, the focus on the 'meaning' aspects of tasting (whether physiological or spiritual) in this demonstration was primarily motivated by a naïve and practical perspective on maximising educational impacts, not by some theoretical perspectives (such as Deweyan educational theory). Although not directly addressed above, some children were suspicious of the use of artisanal products in the educational intervention. This usage provoked their questions such as 'industrial soy sauce actually tastes better and can be produced cheaply, so why not?' and 'after all, why are traditional cultures to be preserved?' The content of the current food education programmes, in which the socio-cultural aspects of eating have been largely ignored, would be unable to provide answers to these questions. In this sense, the evidence-based approach is not merely the so-called PDCA cycle, but rather a theoretically consistent evolution; that will eventually justify the following analysis of the cultural and ethical aspects of food education by going beyond its bio-psycho theoretical foundations.

Notes

1 This chapter was reprinted with permission from Ueda, H. (2017) Impact evaluation of a school-based food education intervention on cognitive and attitudinal aspects of children's dietary habits. *Journal of Food System Research*, 24(1), 2–17, and Ueda, H. (2018) Impact evaluation of the updated educational programme of taste education

with its evaluation system improved based on the development of 'food curiosity' scale. *Journal of Japanese Society of Shokuiku*, 12(2), 109–124.

2 Probably due to his professional training (French cuisine), the chef-teacher did not mention *umami* for this programme, so the score was calculated based on the other four basic tastes.

3 When returning to the study of Gaignaire et al. (2011), their definition of food curiosity, without their being conscious of this distinction, mistakenly referred to its trait aspect, while still adhering to its epistemic and specific aspect.

4 Physiologically speaking, *sweet* usually indicates energy-rich nutrients; *umami* is stimulated by the taste of amino acids that constitute muscles; *salty* signifies modulating the diet for electrolyte balance; *sour* and *bitter* do not only facilitate physiochemical reactions, but also enable animals to avoid rotten or toxic foods. The eater inherently favours the former three, while acquiring a preference for the latter two through eating experiences (Fushiki et al. 2013).

5 The fact that the control group, by contrast, experienced decreases in frequencies of 'sweet' and 'sour' after the programme strongly indicated the educational effect of the programme.

References

Berlyne, D.E. (1954) A theory of human curiosity. *British Journal of Psychology*, 45, 180–191.

Boyle, G.J. (1979) Delimitation of state-trait curiosity in relation to state anxiety and learning task performance. *Australian Journal of Education*, 23, 70–82.

Fushiki, T., Kawasaki, T. & Manabe, Y. (2013) *Oishisa no seiri to shinri* [The physiology and psychology of palatability]. In Y. Yamano (ed.) *Oishisa no Kagaku Jiten [Dictionary of Sciences of Palatability]*, 1–18, Asakura Shoten.

Gaignaire, A., Politzer, N., Monnery-Patris, S., Delaroche-Houot, S. & Dubois, D. (2011) *Evaluation des effets des Classes du Goût*. Ministère de l'Agriculture de l'Agroalimentaire et de la Forêt.

Guyatt, G.H. (1991) Evidence-based medicine. *American College of Physicians Journal Club*, 114, A-16.

Hayabuchi, H. (2015) *Mikaku no jyugyo no koka kensho [Educational effects of 'Leçon de Goût']*. Oral presentation at the *Sekai no kyosho tono deai [Meeting with world master chefs]* Symposium, September, Tokyo, Japan.

Hayabuchi, H., Ueda, H., Umeki, Y., Eto, W. & Ota, M. (2017) Effects of the 'Lesson of Taste' on taste recognition, food awareness, and eating behaviors among elementary schoolchildren. *Journal of Japanese Society of Shokuiku*, 11(4), 323–333.

Huis, A., Itterbeeck, J., Klunder, H., Mertens, E., Halloran, A., Muir, G. & Vantomme, P. (2013). Edible insects: Future prospects for food and feed security. *FAO Forestry Paper*, 171, 1–191.

Imada, S. & Yoneyama, R. (1998) Psychological studies about eating behavior (4): A Japanese version of the food neophobia scale (FNS). *Studies in the Humanities and Science*, 38(2), 493–507.

Kubota, K., Uchisaka, Y. & Tokyo Gas Ltd. (2011) *The Guideline for Leçon de Goût*. La Semaine du Goût Japan Office.

Leherissey, L. (1971) *The Development of a Measure of State Epistemic Curiosity*. Florida State University, Ph.D. Thesis.

Loewenstein, G. (1994) The psychology of curiosity: A review and reinterpretation. *Psychological Bulletin*, 116(1), 75–98.

Mustonen, S., Rantanen, R. & Tuorila, H. (2009) Effect of sensory education on school children's food perception: A 2-year follow-up study. *Food Quality and Preference*, 20, 230–240.

Naylor, F.D. (1981) A state-trait curiosity inventory. *Australian Psychologist*, 16(2), 172–183.

Park, B.K. & Cho, M.S. (2016) Taste education reduces food neophobia and increases to try novel foods in school children. *Nutrition Research Practice*, 10, 221–228.

Pilner, P. & Salvy, S.J. (1971) Food neophobia in humans. In R. Shepherd & Raats, M. (eds.) *The Psychology of Food Choice*, 75–92. CABI Publishing.

Reverdy, C. (2011) Sensory education: French perspectives. In V.R. Preedy, Watson, R.R. & Martin, C.R. (eds.) *Handbook of Behavior, Food and Nutrition*, 143–157. Springer.

Reverdy, C., Chesnel, F., Schlich, P., Köster, E.P. & Lange, C. (2008) Effect of sensory education on willingness to taste novel food in children. *Appetite*, 51, 156–165.

Reverdy, C., Lange, C. & Schlich, P. (2004) A sense of identity. *Oral Presentation at 1st European Conference on Sensory Science of Food and Beverages*, September, Florence, Italy.

Reverdy, C., Schlich, P., Koster, E.P., Ginon, E. & Lange, C. (2010) Effect of sensory education on food preferences in children. *Food Quality and Preference*, 21, 794–804.

Seto, K., Kusumi, T., Tsujimoto, C., Yamamoto, T. & Sawai, S. (2005) *Aji Kotoba no Sekai [The World of Taste Expression]*. Kaimeisha.

Tokumaru, S., Endo, Y., Aoki, M. & Okui, K. (2014) Practice of the taste educational program in junior high school home economics course. *Bulletin of Joetsu University of Education*, 33, 205–216.

Tsuji, M., Tsunoda, M., Suzuki, R., Ikeuchi, R. & Fukushima, T. (2011) The food and nutrition education in Japan for the current problems in food and nutrition. *Fukushima Medical Journal*, 62(2), 79–87.

Ueda, H. (2017) Impact evaluation of a school-based food education intervention on cognitive and attitudinal aspects of children's dietary habits: A case study of a sensory education program 'Leçon de Goût'. *Journal of Food System Research*, 24(1), 2–17.

——— (2018) Impact evaluation of the updated educational programme of taste education with its evaluation system improved based on the development of 'food curiosity' scale. *Journal of Japanese Society of Shokuiku*, 12(2), 109–124.

5 Institutionalisation in Japan, France and Italy

In this chapter the analytical focus shifts to the institutional dimensions of taste education. As noted in Chapter 2, most food education programmes tend to be transient. Therefore, an institutional analysis, with a view to establishing its sustainable promotion, constitutes the second thematic group of food education research. Taste education is a promising case because it can bring a unique perspective grounded in nearly 50 years of experience, having been around since the 1970s. It also enables us to perform international comparative analysis (France, Italy and Japan) to create universal implications for a more sustainable institutionalisation. Thus, using health promotion theories in public health, each country's system of promoting taste education is examined in this chapter, with a special focus on (i) its political positioning (ii), the organisational structures of promoter-actors and (iii) their effective strategies within institutions and communities.[1]

5.1 Health Promotion Theories

5.1.1 *Multiple Levels of Influence on Health*

I employ health promotion theories, which are widely used in public health (Glanz & Rimer 2005), to establish the analytical focus for taste education. Health promotion is defined as 'the process of enabling people to increase control over their health and its determinants, and thereby improve their health' in WHO's Bangkok Charter enacted in 2005. It is founded on the ecological perspective, in which the multiple levels of influence on health and their reciprocal causation are emphasised. As such, McLeroy et al. (1988) identify five major levels of influence: intrapersonal, interpersonal, organisational or institutional, community and public policy factors. The first two levels refer to the pedagogical interactions between children and educators (as discussed in Chapters 3 and 4) and are thus omitted from the current analysis.

I first analyse public policy factors, particularly in terms of how taste education is politically positioned in primary school education and how it is actually conducted. Second, I move on to examining the organisational factors of each promoter-actor, such as legal status, structural size and constituent members, organisational functions, professions, decision-making processes and activities.

DOI: 10.4324/9781003341895-7

Third, from the perspectives of institutional and community factors, I analyse their strategies for promoting school-based taste education projects by employing the following two conceptual frameworks.

5.1.2 Community Organising Theory and EPODE Approach

Two theories are employed, according to the actors' functions (which is identified in the second analysis): Community organising theory is used for regional and independent actors who function within or adjacent to concerned communities, whereas the *Ensemble Prévenons l'Obésité Des Enfants* (EPODE: Let's Prevent Child Obesity Together) approach, which is a large-scale community-based health promotion approach, is applied to the central actors who coordinate these regional organisations.

Community organising is a process through which community groups are helped to identify common problems, mobilise resources and develop and implement strategies to attain collective goals (Glanz & Rimer 2005). Although several change models can be utilised, the following five factors have been identified as key to achieving and measuring change (Minkler & Wallerstein 2012): (i) *participation and relevance* (community organising that 'starts where the people are' and is not the initiative of external experts); (ii) *critical consciousness* (a process of making consciousness social, economic and political factors that limit people's lives and the potential for social change); (iii) *issue selection* (identification of winnable and specific targets of change that can unify and build up community members); (iv) *community capacity* (community characteristics that affect the feasibility to address problems and that can be developed through education and training); and (v) *empowerment* (a social action process through which people gain control over their lives and their communities). These factors are used to analyse each taste education project.

On the other hand, the EPODE approach can provide a useful lens for analysing large-scale community-based projects. It was developed from the experience of a French long-term school-based nutrition education programme, the *Fleurbaix Laventie Ville Santé* (FLVS) study, which was followed by a number of intervention projects over 12 years that resulted in a substantial decrease in the prevalence of overweight children (Borys et al. 2012). The EPODE approach became a foundation for planning and measuring large-scale community-based health promotion programmes and has subsequently been applied to 30 programmes in 22 European countries (with the most in France), (EPODE International Network n.d.). Because of its high affinity with the French regional context, as well as having received increasing attention from taste education practitioners (Reverdy 2011), it seemed reasonable to utilise the EPODE approach in this study.

To put the EPODE approach into practice, a central coordination team (CCT) uses organisational techniques to train and coach local project managers in each community. The local project manager's role is to mobilise diverse local stakeholders to foster health-related behaviours among children and families. Using this logic, the EPODE approach can be applied here by positioning central actors as CCTs and regional actors as local project managers. In this chapter, CCTs refer to the central actors who promote taste education, whereas regional and independent actors represent local project managers.

In the EPODE approach, the following four factors are identified: (i) *political commitment* (officially gained from leaders of governments and local authorities); (ii) *resources* (secured at local and national levels to fund central support services and evaluation, as well as local implementation); (iii) *support services* (including partnership management, training programs and other support services for community practitioners); and (iv) *evidence* (to evaluate programmes' processes and outcomes). These factors are used here as a lens with which to evaluate taste education projects.

The data used for analysis included the relevant literature, interviews with the leaders of 11 organisations (France: n = 3, Italy: n = 3, Japan: n = 5), and observations of their organisational activities (classes and training programmes for potential educators). The fieldwork in three countries was conducted from June 2015 to October 2016.

5.2 Political Positioning in Primary Education

I first discuss how taste education is politically positioned in primary education. In all three countries, taste education is positioned as 'one type of food education pedagogy'; thus, its institutional status automatically follows that of food education policies.

5.2.1 Basic Law on Food Education in Japan

Of the three countries, Japan is the only one with a specialised law for food education, called the Basic Law on Food Education. I discussed in Chapter 1 the social and academic context until the enactment of the Basic Law in 2005, so I focus here on its legal nature and the impact on the institutionalisation of taste education. A basic law (*kihon-ho*) is a legal system unique to Japan, so it would be useful to specify its general characteristics (Shiono 2008): (i) It is 'enlightening' but it does not stipulate citizens' concrete rights or duties. (ii) It provides policy ideas and goals, but further plans shall be articulated in subsequent regular laws. (iii) It imposes related policy development on its government. (iv) It is enacted in inter-ministerial ways. (v) It provides few legal rules (often without stipulating rights, duties and sanctions).[2]

In this way, the Basic Law on Food Education prescribes its basic ideas (Article 2–8, the nuances of which are more clearly described in its Preamble), as well as the responsibilities of those concerned (Art. 9–13), one of which applies to those concerned with school education (Art. 11). It also stipulates that the state, prefectural and local authorities shall 'endeavour' to develop basic food education plans (Art. 16–18). It also specifies certain contents of these basic plans, such as the development of food education guidelines and the deployment of nutrition teachers in schools and (Art. 19–25). This Basic Law was followed by the subsequent amendment of related regular laws: the Revised Curriculum Guidelines (2008), in which 'food education' was clearly prescribed as the theme of the relevant subjects (e.g., physical education and home economics), as well as the amended School Lunch Program Act (2009) that requires school meal actors' endeavours to promote food education.

Having outlined the political framework of food education, I highlight its advantage and disadvantage in promoting taste education. The major advantage concerns the basic idea and institutional stipulation that food education is a 'comprehensive' approach to a wide range of food-related issues, as recognised in the Basic Law's preamble (health, food safety, food culture, environment, morality, etc.), and this approach is even proudly announced as 'unique to Japanese *Shokuiku*, whereas in the West, more limited approaches such as nutrition education programmes [...] are practised' (MAFF 2005, p.4). Yet, without any theoretical justification, the stipulated comprehensiveness results in the conceptual ambiguity of food education and a lack of effective political orientation (as seen in Chapter 2). By referring to 'comprehensive' approaches, these political frameworks do not favour nor disfavour any particular pedagogy; concrete pedagogies are to be determined by individual local authorities, schools and teachers. As a consequence, taste education is reported to have been currently practised not only with less frequency than other dominant (mostly nutritional) pedagogies, but also in disciplinarily confined manner, that is, mainly in home economics classes (Sato 2009; Sato & Ishii 2014; Tokumaru et al. 2014), with a few exceptions, namely in integrated studies (Ueda 2017), during lunchtimes (Ueda et al. 2015; Ueda 2018a) and in extra-curricular classes (Sasaki 2008). In sum, despite the establishment of the Basic Law, the current public policy in Japan lacks a concrete orientation to facilitate the envisioned 'comprehensive' approach to food (in general) and the recognition of the promising potential of taste education for this purpose (in particular).

5.2.2 Trans-Ministerial and Transdisciplinary Positioning in France

In contrast to its weak political presence in Japan, taste education is clearly positioned as a recommended pedagogy in two French public policies, *Programme National Nutrition Santé* (PNNS: National Nutrition and Health Program) and *Programme National de l'Alimentation* (PNA; National Food Program). PNNS has been practised as a five-year health promotion plan that was formulated in 2001 and has integrated food education (more precisely 'education [...] through nutritional recommendations') as one of its core objectives (Public Health Code Article L.3231-1). The most recent plan (2019–2023) also positions food education as one of 10 priority objectives, in which taste education is presented as a recommended pedagogy (MSS 2019). On the other hand, PNA is a renewable five-year food policy that has been in operation since 2011. It was institutionalised under the LAAF (Law of the Future for Agriculture, Food and Forest) in 2014, in which the implementation of 'food education for the young' is stipulated as one of the core objectives (Rural Code Article 1). The subtle difference to PNNA is that, in the most recent plan (2019–2023), 'taste education' is defined with a more political importance as one of three priority objectives (MAAF 2019).

Keeping these two political frameworks in mind, I now turn to primary education. The first appearance of taste education in legal texts can be traced back to a circular (No. 118) of the Ministry of Education in 2001, which was aimed mainly at reconsidering the safety standard of school meals (as a counter-movement to BSE issues) but simultaneously promoted taste education as a pedagogy for learning the cultural values of food as a complement to nutrition education. In accordance with the progress of PNA and PNNS, a circular (No. 216) in 2011 announced the orientation of health education, in which 'nutrition education' was regarded as one of the major areas to combat child obesity and 'taste education' as one of the recommended pedagogies. In 2014, when the LAAF was enacted, the implementation of 'food education' in schools was finally stipulated in the Education Code (L312–17), consistently with PNA and PNNS. The Modernisation of Health Systems Law in 2016 (Education Code L.541-1) further reinforced the institutionalisation of school-based health education by revalorising the *Parcours Educatifs de Santé* (PES: a health education course that was developed in the Vincent Peillon Law in 2013), in which subjects related to food (such as nutrition and digestion) are included and thus considered a part of 'food education'.

In sum, food education is legally stipulated in the Public Health Code, the Rural Code and the Education Code, the third of which stipulates its implementation in primary education. Moreover, taste education is clearly situated as a recommended pedagogy or priority objective in two corresponding public policies (PNNS and PNA). (In addition, I later discuss its positioning from the perspective of cultural policy.) Perhaps due to the success of such a diverse political justification in France, taste education is reported to have been a transdisciplinary practise in a wide range of national curricula, such as language, discovery of the world, science, social studies (Togawa 2009) and other extracurricular activities, such as during lunchtimes (Gaignaire & Politzer 2010).

5.2.3 *Increasing Institutionalisation in Italy*

In contrast to Japan and France, food education in Italian schools has yet to obtain legal status, but it has been gradually institutionalised. According to Strozzi (2015), 'nutrition education' in schools has attracted social attention since the 1970s and its scientification and institutionalisation started in the 1980s. Diverse food pedagogies for schools have been developed in Italy with the active participation of food companies and Slow Food activists since the 1990s, and in the EU since the 2000s. Yet, it has been unable to obtain political commitment and has remained weakly institutionalised. Reflection on the limits of traditional nutrition education finally started in 2011, when the Scientific and Technical Committee on Schools and Food (under the Ministry of Education) was established and the Guideline on Food Education for Italian Schools was developed based on the committee's proposals. The guideline was revised in 2015 when an expo for food cultures took place in Milan.

Although food education in primary education has been gradually institutionalised in Italy, there are no equivalent legal frameworks for food education in general and taste education in particular that are equivalent to those of France and Japan. Nevertheless, it is interesting to note that the revised guideline of 2015 promotes a 'systematic approach' as a reflection on its poor institutionalisation and it then situates taste education as a promising alternative (p.21). This guideline also requires the transdisciplinary practice of food education in schools, but further studies are expected to elucidate the actual practices in Italian schools.

5.3 Organisational Structures of Promoter Institutions

The interviewed promoter institutions were categorised according to their organisational functions: (1) Regional or independent organisations that are working within the community, namely the Food Consciousness Institute (FCI), Japan Taste Education Association (MIIKU), Slow Food Suginami TOKYO (SF Tokyo), the Institute of Children's Taste Development (IDGE), *Eveil'O'Goût* and the Italian Taste Education Centre (CEG); and (2) central institutions that unite these regional organisations, namely the *La Semaine du Goût* (LSG) offices in France and Japan, Slow Food Japan, the National Association for Taste Education for the Young Association (ANEGJ) and Slow Food International (SF International). Note that another important promoter organisation in France, *Institut du Goût* (Institute of Taste), is not included here but its function is discussed later in this chapter.

The other elements that define their organisational structures are summarised in Table 5.1, but I should highlight that the individual efforts of regional organisations have serious limitations concerning their organisational sizes (two to six staff members), professions (specialising more in pedagogies than in promotion) and educational opportunities (e.g., IDGE: 10 classes per year, SF Tokyo: 30 students per year), which thus necessitates support services from central organisations.

5.4 Community Organising of Taste Education Projects

To provide a vivid image of community-level practices, I detail the most characteristic projects of the selected three organisations (Italy: CEG; France: *Eveil'O'Goût*; Japan: SF Tokyo) in relation to the six factors proposed in the community organising theory. The data for analysis includes the literature on each project (Sasaki 2008; CEG & Nakano 2012; Gaignaire & Bourgeois 2016) and additional interviews with the leaders of each organisation.

5.4.1 A *Scuola con Gusto* in Italy

Viareggio is a port town in the northwest of Tuscany and fishing constitutes one of its primary economic sectors. Despite having a food culture that is unique to the fishing village, one of the major challenges for the city government's educational division was the vast amount of food leftovers from school meals. The CEG was consulted by the educational division and it then created opportunities

Table 5.1 Organisational Structures of Promoter Institutions

Organisation (since)	*Statue*	*Size*	*Function*	*Profession of staff*	*Activities*	*Pedagogy*
LSG Japan (2010)	Ltd.	2–5 staff members	Central	Management	Classes (in schools) Training programmes	LSG
FCI (2011)	Association	6 directors 2–3 staff members	Independent	Education Ecology Business	Classes (outside schools) Training programmes	*Shinagawa*
MIIKU (2011)	Association	2 staff members	Independent	Education Cookery science	Classes (outside schools) Training programmes	MIIKU
SL Japan (2004)	Association	3–5 staff	Central	Management	Classes (outside schools)	Slow Food
SL Tokyo (2002)	Voluntary	2–5 staff members	Regional	Education	Classes (in schools)	Slow Food
IDGE (2012)	Voluntary	6 directors 2 staff members	Independent	Education Home economics Physiology of taste	Classes (in schools) Training programmes	*Puisais*
LSG France (-)	Ltd.	20 staff members	Central	Management	Classes (in schools)	LSG
ANEGJ (2012)	Association	10 directors 2–3 staff members (25 regional associations)	Central	Education	Training programmes	*Puisais*
Eveil'O'Goût (2010)	Association	3 directors 3–5 staff members	Regional	Education Behavioural science	Classes (in schools) Training programmes	*Puisais*
CEG (2008)	Association	2 staff members	Independent	Education Local history	Classes (in schools) Training programmes	Prato
SL International (1989)	Association	(360 regional associations)	Central	Management	Classes (outside schools)	Slow Food

for dialogue with five kindergartens (and later elementary schools), a catering service company and producers in the community (*participation*). In the ensuing dialogues, these participants discussed why the children did not like the taste of the school meals and eventually realised the need for community members (including the children themselves) to 're-evaluate the quality of food, their own local traditions and cultural heritages' (*critical consciousness*). According to the CEG's proposition, these members then decided to run a multi-year taste education project in schools to address the challenge (*issue selection*).

In the first year, the CEG organised a series of workshops, in which the findings of a survey on the children's eating habits were shared with their parents and training programmes were conducted for potential taste educators, including parents, teachers and catering staff (*community capacity*). Following the success of the second year's project, the third year saw the construction of a school garden to stimulate the children's curiosity about vegetables and further foster community members' engagement (*participation*). In the fourth year, the improvement of school meals finally began, based on the previous years' achievements. This required a dual process of developing menus using local products (notably fresh fish) to reflect the identified strength of Viareggio and of mobilising the previously trained educators to implement taste education workshops to reduce the children's food neophobia towards the newly integrated products (*community capacity*). All of these processes led not only to the reduction of leftovers (the initial objective), but also to the establishment of a trust relationship between schools and producers, which facilitated regular farm visits as school programmes (*empowerment*).

5.4.2 BioSenso in France

Dijon is a city in the *Bourgone-Franche-Compté* region. It is a major academic city, in which there are some institutions that specialise in food and wine research, such as the University of Bourgogne and Centre for Taste and Feeding Behaviour (CSGA). Dijon also serves as one of the gastronomic towns. It is thus no surprise that *Eveil'O'Goût* was established in such a gastronomic and scientific atmosphere, with the purpose of stopping an increasing 'impoverishment' of children's taste. To achieve this mission, *Eveil'O'Goût* organised a two-month taste education project called '*BioSenso*', with financial support from an agri-food foundation. *Eveil'O'Goût* shared this mission with relevant institutions (*participation*) and two after-school activity centres in the community eventually expressed interest in implementing this taste education project (*issue selection*).

The project consisted of a series of taste education classes using organic and alternative products. The project took place mainly as an after-school activity, but the sustainable recipes developed by the participants themselves based on their learning in the project were enjoyed during lunchtimes by 1,300 school children in the same community. The unique aspects of *BioSenso* were the process of developing the project content. Despite the potential difficulty of teaching young children about sustainability and various food labels, *Eveil'O'Goût* mobilised scientific evidence on children's eating behaviours from the CSGA to define its educational content and methods, such as the use of puppets and ludic plays (*community capacity*). The project's success enabled *Eveil'O'Goût* and the activity centres to develop a rapport, and these community members became more confident and motivated to launch future taste education projects (*empowerment*) (Figure 5.1).

5.4.3 *Finding Real Tastes in Japan*

Suginami is a ward in Tokyo that is well-known for its urban agriculture. Despite its community strength, SF Tokyo has worried about 'an increasing distance between farmers and consumers'. SF Tokyo accessed the community's parent-teacher association (PTA) to raise parents' awareness of the issues (*participation*). In their dialogues, these members eventually recognised the limits of traditional nutrition education and home economics, and the need for a novel pedagogy centred on 'taste' and 'pleasure', as well as community collaboration to implement this type of educational content (*critical consciousness*). To address these challenges, SF Tokyo reached out to merchants in shopping streets, local farmers and chefs in the community, and finally decided to implement a series of taste education classes as an optional Saturday class at school (*issue selection*).

The project involved procuring food from local farmers and mobilising chefs from local restaurants (*community capacity*). Although being infrequent (once per academic semester), a constant practice of taste education classes strengthened the relationships between the PTA, the farmers and the professionals in the shopping streets, which paved the way for other projects, such as a community café for children in need (*empowerment*).

5.4.4 *Effective Strategies for Community Organising*

Drawing on these three cases, some effective strategies can be identified. The first relates to the range of empowerment which, in theory, represents change processes both for individuals and the community's social structure (Zimmerman

Figure 5.1 The *BioSenso* Project in France

Source: Photographed by the author during the observation. The pictures are of an educational programme for kindergarten children and various food labels that are used as educational material.

2000). In the French and Japanese cases, empowerment occurred mainly between the individuals directly involved in the projects. On the other hand, the Italian case entailed the transformation of the community's social structure, notably through school gardens and the local food supply chain being channelled into school meals, which would increase the sustainability of the projects.

Second, the participation of various stakeholders was identified as a prerequisite for community empowerment. Local food professionals can provide not only technical and physical resources (as exemplified in the Japanese case), but also justify their participation on the basis of 'preserving local traditions', thus avoiding possible criticism about their commercial motivations, which have often been directed at big food companies (Takano 2012). Parents are another key set of stakeholders, but their participation has often been problematised in the food education literature. Although there is no panacea, in the Italian case the parents' participation was effectively encouraged through 'strategically devised dialogues' that integrated the evidence-sharing regarding their children's eating and health, and the training programmes for developing their own tasting abilities.

Third, the other effective strategy, which was employed in the French case, involved collaboration with scientific institutions (the CSGA) to give their project impetus, the importance of which is briefly revisited in the next section and again in more depth in the last section on the historical development of taste education in France.

5.5 Effective Strategies of Central Organisations

To identify how central organisations can help regional or independent actors to overcome their structural limits and bring impetus to their community-based projects, the five central organisations (the LSG offices in Japan and France, ANEGJ, SF Japan and SF International) were analysed on the basis of four effective factors informed by the EPODE approach.

5.5.1 *Political Commitment*

It is particularly notable that four of the above organisations (excluding SF Japan) were formally recognised by the ministries concerned, which helped them to gain social credibility and facilitate further promotion of their projects. Another characteristic is that the Slow Food-related central organisations established partnerships with the local authorities, such as SF International's partnership with the City of Turin for the bi-annual organisation of *Terra Madre* (one of the largest agri-food festivals in the world, in which taste education classes are performed) and SF Japan's collaboration with the City of Kobe in Japan to make it a 'Gastropolis', in which taste education is positioned as one of the priority objectives.

5.5.2 *Resources*

Financial support from private food companies has merits and disadvantages. It creates impetus for organisational activities, but it also risks the integrity of educational projects. This complexity was clearly illustrated by the LSG offices. The French LSG has been heavily criticised for its initial relationship with the Sugar Collective, which aimed to justify 'sweet' as one of the five basic tastes and thereby increase the consumption of sugar. The Japanese LSG has been largely dependent on the food companies' funding and, for example, the use of donated 'artificial candy' for educational purposes has been controversial. To avoid these problems, the EPODE approach suggests confirming in advance that private funders shall not interfere with a project's educational content; however, doing so may, paradoxically, decrease the private sector's motivation to offer financial support.[3]

One solution is to prioritise relationships with (inter-)professional organisations, which, in turn, coordinate the collective interests of constituting associations or business operators involved with their common products (Niiyama et al. 2014). They will, thus, be more committed to social issues around their products. Indeed, promoter institutions of taste education in France (including the LSG office) have been supported for over 20 years by (inter-)professional organisations, such as the Inter-professional Association of Cattle and Meat (INTERBEV) and Inter-professional Association of Fresh Fruits and Vegetables (INTERFEL). For example, INTERFEL situates food education as 'their obligation to citizens, not brand image' and mobilises about 20 professional staff members (including dietitians) to promote food and taste education using fruits and vegetables in schools (Niiyama & Ueda 2020). Moreover, Italian national food education projects such as *Scuola e Cibo* (Schools and Food), which include taste education, have also been supported by inter-professional organisations, such as the Italian Federation of Food Industries (*Feralimentaire*) and the Agricultural Product Confederation (COPAGRI). The social commitment of professional organisations to food education in Europe provides critical insights for Japan, where these entities (particularly agricultural cooperatives) have been criticised as merely promoting their own economic interests.

5.5.3 *Support Services*

The identified effective support services include constructing regular information platforms for regional constituents (ANEGJ and SF International), partnership management (LSG offices with local schools and chefs), and building press relations (all five institutions). Among these, developing training programmes for potential educators has been shown to be particularly important. Some organisations (notably the FCI and ANEGJ) promote the training programmes, not only for the theory-learning part of taste education, but also for leadership development. The latter is a necessary capability for facilitating community members' horizontal dialogue and their critical consciousness (Freire 1968), and is thus consistent with the theory of community organising. As partly described elsewhere, the regional organisations also provide the training programmes, among which those of *Eveil'O'Goût* and the CEG integrate follow-up support for the participants to plan interventions and obtain the necessary funding, rather than merely issuing certifications at the end of the programmes.

5.5.4 Evidence

Among them, the ANEGJ is the most committed to gathering evidence of the educational impacts of taste education, mostly because it is composed of regional organisations that were previously involved in a nationwide taste education project (Schlich et al. 2014). The other organisations only manage to track their project scale (e.g., the number of participants and the national percentage of those who recognise their projects). This does not mean that these organisations are not motivated to conduct scientific studies on the educational impacts, but they are unable to do so due to inadequate resources.

To address this challenge, the first step is to establish basic monitoring and evaluation systems. Concretely put, effective strategies would be to enhance the central organisations' CCT function to collaborate with and collect data from schools, and health and agri-food agencies in the community; to activate its information platform and conduct interviews with regional organisations; and to consult with academic institutions, particularly in the planning stages. Below, I describe the French case to indicate how such evidence has contributed to the institutionalisation of taste education.

5.6 History of Institutionalisation of Taste Education in France

From the perspective of health promotion, I have so far discussed the challenges and effective strategies for promoting taste education. Although it is tempting to apply the same strategies to other regions, the specificity of each project should also be recognised. Health promotion theories did not take us in this direction, so I discuss the historical evolution of taste education in more depth by taking France as the most telling example, with additional evidence from recent fieldwork (2018–2019).

5.6.1 Preparatory Stage (1927–1971)

First of all, it would be useful to understand the life history of Puisais (1927–2020) up to the development of his method in 1972. He was born into a family of wine merchants in Poitiers in 1927. After obtaining a doctoral degree in science, he started his professional career as an analytical chemist-oenologist and took up a position in the Departmental Laboratory of Analysis and Research at Indre-et-Loire in 1959 (Polony 2010). The turning point for Puisais was in 1964, when he attended a scientific conference on food and the senses organised by taste physiologist J. Le Magnen (1916–2002), and was struck by the complexity of taste, merely a part of which was attributable to the physical qualities of food (water, soil, etc.), which he had analysed throughout his career (Politzer [2011]2016). Heading back to the laboratory, he immediately launched a new research project on the connection between food and taste (Cody 2009). It must be noted that Le Magnen, together with nutritionist J. Tremolière (1913–1976), contributed immensely to the establishment of the contemporary physiology of taste and eating behaviours. Puisais pioneeringly developed his pedagogy based on this emerging scientific tradition, but his audience was initially limited to professionals (Politzer 2016).

5.6.2 *Establishment Stage (1972–1998)*

Puisais began directing his pedagogy at children in 1972, when he was consulted by students of sociology at the University of Tours and launched an experiment on the connections of children's (aged 10–11) olfactory systems and pedagogical influences, after realising that they had keener olfaction than adults but simply lacked the vocabulary to convey their perceptions of such subtle senses (Puisais 1999, pp.36–37). Based on the *Conservatoire du Goût* (Conservatory of Taste), which he had established in the previous year, a programme consisting of 10 sessions (which was later called *Classes du Goût*) was launched in primary schools in the Centre Region (Politzer 2016). The conservatory was upgraded in 1976 to the French Institute of Taste (IFG), in which his pedagogy was further extended through interdisciplinary dialogue to serve as 'a centre of reflection [...] focused on the bio-psycho-socio-economic constituents of taste [...] for the educational purpose of children's human development' (IFG internal document, p.1). This focus on the totality of food was well represented by the composition of its scientific committee (physiologists, a psychologist, a nutritionist, a historian of mentality and food sociologists) as well as by the thematisation of its annual conferences (such as 'Taste and Milieu', 'Taste and Grammar', 'Taste and Imagination' and 'Taste and French Culinary Heritages').

It is of particular importance that 'sociologists of eaters', including C. Fischler and J.P. Corbeau, were also engaged as committee members. To recall the scientific history of sociology (Poulain 2017, p.132), food had long been an ambiguous subject: On the one hand, it was viewed as a physiological object and thus excluded from sociology. On the other hand, eating and drinking, as behavioural patterns, were included as social facts (Durkheim 1895, p.58). To rescue food from such ambiguity, as well as to go beyond the previously narrow focuses in the 'sociology of food consumption' (Halbwachs, etc.) and the 'sociology of taste' (notably Bourdieu), the sociology of eaters emerged in the 1970s from the Maussian tradition, in which food was viewed as a total social fact (Mauss 1968) and from Morin's (1973) transdisciplinary philosophy, as an approach to food and eating as a 'total human act' (Poulain 2002, p.16) at the intersection of bio-psycho-socio-cultural dimensions. Sociologists of eaters have contributed to articulating the totality of food-related subjects, which were previously reduced to single disciplines, such as taste and gastronomy (Poulain 1985, 2017), gastro-anomy (Fischler 1990), sociability (Corbeau 1997), obesity (Poulain 2009) and pleasure (Dupuy 2013). Through both professional and personal interactions with these scholars, it is impossible to deny Puisais' transdisciplinary thematisation of taste in the IFG's organisational activities.

Taste education has been socially recognised since the publication of the first textbook, *Le Goût et l'Enfant*, on the basis of the IFG's 10 years of experience of taste education practices. This social movement led to the *La Journée du Goût* (The Day of Taste) initiative in 1990 and its expansion to LSG in 1992, as well as to the establishment of the National Council of Culinary Arts (CNAC) in 1990, the role of which included promoting this national movement. The CNAC is of particular importance, in the sense that it highlights a new positioning of taste education. This entity was established with the purpose of promoting the French culinary heritage and taste education under Jack Lang's cultural politics (1981–1993), which originated from the birth of the Socialist Party's regime in 1981 and gave impetus to a wide range of cultural activities (including popular cultures) and their institutionalisation (school-based education, the development of cultural professions, etc.) (Girard 2009). Although the CNAC was, unfortunately, dissolved due to financial problems, its many years of support for food education paved the way for the IFG to continue its organisational activities. By 1999, the IFG had reached 100,000 children in 3,000 classes and 250 educators (school teachers, school nurses, etc.), who took part in their training programmes (Puisais 1999, p.132).

In sum, the theoretical foundation of Puisais' pedagogy was already established at this stage, mainly through the newly established physiology of taste (Le Magnen and Tremolière) and partly through interdisciplinary dialogue, most notably with the food sociologists. On the other hand, taste education was increasingly institutionalised due to political support, which situated it as covering the 'cultural' dimension of food in contrast to traditional nutrition education.

5.6.3 *Incubation Stage (1999–2009)*

The incubation stage was characterised by the expanding connection, not only with the Ministry of Culture, but also with the ministries of education and agriculture, as well as evidence-building on its educational impacts on a national scale, both of which prepared the way for the succeeding development of taste education. At the organisational level, Puisais created the *Institut du Goût* in 1999 and merged the IFG into this new entity two years later. Although its activities were modest due to the aforementioned dissolution of the CNAC, the institute restarted taste education practices in 2001 within the Artistic and Cultural Project (PCA), a new political framework initiated by the ministries of culture and education. More concretely, the institute constructed a new domain of 'Art of Taste' in the database of educational materials in collaboration with the local educational authority in Poitiers (Politzer 2004). This case demonstrated the growing interest of the Ministry of Education in taste education, which is consistent with the first appearance of the term 'taste education' in legal texts in June 2001 (in circular No. 118, as discussed above). In 2002, *Classes du Goût* saw the programme expand from 10 to 12 sessions, which dealt with 'food consumption' and 'food labels' (Politzer 2004). However, the termination of the PCA again in 2003 resulted in a resource shortage that affected the institute's continuous activities (Politzer 2016).

Nevertheless, the political interest in taste education has continued to grow. In 2002, in partnership with the Institute of Taste, the Ministry of Education conducted a survey on the actuality of taste education being widely promoted all over France. In 2005, in collaboration with the National Institute of Agronomy Research (INRA), the ministry also launched a national-scale survey on the educational impacts, called EduSens, which proved that taste education could reduce children's food neophobia, improve their taste expression abilities and enable them to form preferences for complex flavours (Schlich et al. 2014). As suggested by the EPODE approach, this EduSens project constituted important evidence that the political intervention in taste education was justifiable.

In addition, taste education also started to occupy a unique space in the food education domain. In 2008, the French Institute of Nutrition (IFN) organised a conference titled 'From Nutrition Education to Taste Education' and proposed a clear distinction between these two pedagogies. Taste education was defined as a 'non-normative' approach based on 'sensory experiences', as opposed to the normative approach of nutrition education, which was based on scientific knowledge of good and bad food (IFN 2008). Meanwhile, the Ministry of Agriculture finally institutionalised its interest in taste education by setting up a special task force (MAAF 2012), which continued the previous study of the Ministry of Education and conducted an additional survey. It also elucidated the unexpected expansion and diversity of taste education practices, as well as the presence of some practices which were disconnected from Puisais' original intention (Politzer 2016). This realisation led the Ministry of Agriculture to construct a national network for taste education actors and to articulate the pedagogical identity of taste education in the next stage.

In sum, the incubation stage of taste education was characterised by three social phenomena. First, its 'politicisation' was accelerated by the engagement of the Ministry of Culture, as well as the ministries of education and agriculture. Second, its 'scientification' was completed by various scientific actors, such as the INRA and IFN. Third, as clearly announced in the IFN's proposal, its 'distinction' from nutrition education was recognised. Although these three phenomena certainly gave impetus to the institutionalisation of taste education, the latter has gradually been detached from its founder's original intention.

5.6.4 *Development Stage (2010–2020)*

An increasing politicisation of taste education sharply marked this development stage, beginning with a couple of political events in 2010. In July of that year, the enactment of the Law on Modernisation of Agriculture and Fisheries ignited discussion on the PNA, in which taste education was situated as one of the priority policies and was set to be promoted at schools in relation to the PNNS. Moreover, the task force under the Ministry of Agriculture developed the 'reference' that defined the objectives of taste education and the necessary qualities of educators (MAAF & MNE 2011). In November, UNESCO inscribed the 'gastronomic meal of the French' as an intangible cultural heritage. Taste education thus acquired force as one of its major safeguarding policies (Billiaux 2016, p.32). Here, taste education was again associated with the 'cultural' dimension of food and eating.

The national expansion of taste education proceeded in the following phases: During the 2010–2011 semester, taste education classes (a shortened version of eight sessions) were conducted in 6 regions by previously trained educators and 15 newly trained school teachers, which reached over 2,000 fifth-grade children (Gaignaire et al. 2011, p.6; Dupouy 2012). During the same period, the Ministry of Agriculture organised a two-day conference dedicated to taste education actors in January 2011, at which the construction of its national network was formally announced (MAAF 2012). In September, the new programme of eight sessions, which had been used for the first stage, when its practical feasibility had been demonstrated, was developed by the ministries of agriculture and education (Montoux et al. 2012). Responding to the 'heritisation' of gastronomy and the new engagement of the National Institute of Origin and Quality (INAO), this new programme included educational content about 'culinary heritages' and 'food labels' in line with the political intention to increase the consumption of alternative and labelled products (MAAF 2011). This pedagogical revision has been a concern for taste education actors who aim to protect its educational integrity and deny its 'objective of promoting some particular food products' (Billiaux 2016, p.225). In any case, this programme became the major educational material to be promoted widely in all the French primary schools.

The national network for taste education actors was finally established in October with the support of the Ministry of Agriculture and the agri-food cluster (Vitagora), which later became the association called ANEGJ (MAAF 2012). This central organisation initially consisted of about 80 regional organisations but, due to some organisational problems, in 2016 its size was reduced to 20 constituents, which weakened its organisational function (Billiaux 2016, p.37). Therefore, as I discuss below, the current promotion of taste education is more subject to local authorities than this national network. Another turning point was seen in November 2011 when the General Directorate of Food in the Ministry of Agriculture organised a four-day training programme for referential educators, in which, again, not only physiologists and psychologists of food but also sociologist of eaters (J.P. Poulain) were engaged. Since this production of referential educators, the initiative has been gradually transferred from the central (General Directorate) to the Regional Directorate of Food, Agriculture and Forestry (DRAAF). Under these regional initiatives, the second large-scale survey was conducted in 25 regions, as well as in some overseas prefectures for the following semester (2011–2012), which again proved the educational impacts on school children (Gaignaire et al. 2011).

As noted earlier in this chapter, the subsequent politicisation at the national level (such as PNA, PNNS and PES) certainly contributed to developing its institutionalisation, but, at the present time, the actual implementation is heavily dependent on the local authorities. Lastly, I take the Occitanie region as the most telling example to demonstrate the actuality of such territorialisation.

5.6.5 Territorialisation: The Case of the Occitanie Region

Located in the southwest of France, Occitanie is the administrative region comprising the previous Midi-Pyrénées and Languedoc-Roussillon regions, which were merged in 2014. I first highlight some of their characteristics based on Billiaux's (2016) helpful survey on the actuality of taste education practices being promoted by the DRAAF in these regions (Table 5.2).

In both regions, the public officials in charge of the PNA in the DRAAF were automatically responsible for taste education, the characteristics of which defined its territorialising approaches. In the Midi-Pyrénées region, DRAAF activated the previously-formed partnership with the local health authority, appointed two nutritionists as referential educators and conducted a total of seven programmes of taste education in the Lot department during one semester (2011–2022). This effective partnership was facilitated by their recognition of 'complementarity' between taste education (DRAAF) and nutrition education (health authority). Nevertheless, to build a more sustainable promotion system, the DRAAF later favoured reinforcing the partnership with the local education authority because it was school nurses (not nutritionists) who were present at the schools and thus more flexible for taste education's implementation. The previous referential nutritionists were replaced with three new school nurses, who made considerable efforts to train 144 school nurses as taste educators until 2015. In this sense, taste education in the Midi-Pyrénées region was territorialised more as a 'health education' approach.

On the other hand, in the Languedoc-Roussillon region, the DRAAF was unable to build a partnership with the health authority, but the impetus for territorialisation came from the official who was in charge of 'Art of Taste' (which was a part of cultural and artistic education) in the local education authority and, interestingly, also served as a referential educator. Thanks to the charisma of this

Table 5.2 Taste Education in Previous Midi-Pyrénées and Languedoc-Roussillon Regions

	Midi-Pyrénées Region	*Languedoc-Roussillon Region*
In charge of PNA	Multiple staff (only one staff member since 2014)	Two staff members (one of which specialised in taste education)
Partnership	Health authority (only 2011–2012) Education authority (only school nurses)	Education authority (the post of 'Art of Taste')
Reference	Two nutritionists (only 2011–2012) Three school nurses (since 2012)	One teacher at hotel school One sensory analysist
Practice (2011)	Seven programmes (all in Lot department)	Four programmes (in four departments)
Trained educators	144 school nurses	178 animators (of which 108 are school teachers)
Other activities	Financial support for taste education projects	Three conferences on taste education (760 audiences) Financial support for 'Art of Taste' Construction of a local network for taste education

official, the DRAAF managed to run each programme in 4 departments and train 178 animators during the 2011–2012 semester, organising 3 conferences dedicated to taste education. The participants, in turn, constructed a local network for taste educators. It can be said that taste education in the Languedoc-Roussillon region attracted recognition as a 'cultural/artistic education'.

The rest of the story was acquired during my interview with the current official responsible for taste education in the DRAAF of the Occitanie region. After the administrative integration in 2014, the DRAAF succeeded in expanding the local network (which had been limited to Languedoc-Roussillon) to the entire region and continued its financial support for taste education projects (10,000 euros per year). However, two challenges emerged with respect to further territorialisation. As is often problematic in any authority, the first is that charismatic official ('Art of Taste') was assigned to another position in a personnel reshuffle, which resulted in the weakened presence of taste education as artistic education in schools. To resolve this, the DRAAF is planning to extend its training programme to potentially include public officials. Another challenge, although not entirely new, is related to the difficulty of ensuring that 8–10 sessions are included in the current school curriculum. The DRAAF is also aiming to train animators who could practise taste education as an extracurricular activity (as already seen in the case of *BioSenso*). This testimony demonstrated the importance of 'training programmes' as one of the support services.

5.6.6 *Voice from Classroom in the Lot Department*

I conclude this analysis by noting the evidence provided by a school nurse, Marianne, who has been actively promoting taste education in the Lot department. In pursuit of innovative food pedagogies, she joined the aforementioned training programme in Midi-Pyrénées in 2011 and, with the help of the local education authority, has trained new school nurses as taste educators every year since 2012.

Drawing on many years of field experience, Marianne identifies four reasons for its effective implementation. The first is relationship-building with homeroom teachers. She had previously developed a rapport with them in the course of jointly teaching normal health education (such as on sexuality, sleep, exercise and food), and as she emphasised, 'taste education is just an extension of such practices'. This field input clearly supports the Midi-Pyrénées DRAAF's shift of focus in the training programmes from dietitians to school nurses. The second reason relates to the small but constant funding from the *coopérative scolaire* (parents' cooperative) of 40 euros per year (which is sufficient to conduct about 6 sessions). This is another way of encouraging parents' participation, which was problematised earlier in this chapter, in taste education projects. The third is the impact of Health Education Guideline (PES), by which the 'consistency' in dealing with food in every subject became authorised and thus helpful for practising taste education. The fourth is its 'systematic' approach (e.g., theoretical basis, a series of 6–10 sessions), which can facilitate obtaining the parents' understanding, as they appraised one day, saying 'that is really a well-designed programme' (Figure 5.2).

Figure 5.2 The Actual Practice of Taste Education in Gramat City of the Lot Department
Source: Photographed by the author during the observation. The photographs are of a school nurse (Marianne) preparing the taste education class and flavoured samples.

5.7 Concluding Remarks

This chapter was aimed at identifying challenges and solutions for building more sustainable systems for promoting food education. To achieve this, the effective strategies at multiple levels (community, organisational and public policies) were analysed from the perspective of health promotion theories, as well as in light of an in-depth historical analysis of the institutionalisation in France. I conclude this chapter by articulating implications for taste education in particular and food education in general.

For the former, three social conditions emerged in France which, on the one hand, gave impetus to the institutionalisation of taste education, but, on the other hand, complicated its educational integrity. A series of new political frameworks (Lang's cultural politics, PNA, PNNS and PES) has increasingly 'politicised' taste education, which used to be a local practice at schools along the Loire River, and was transformed into a pedagogy for stimulating children's consumption of healthy, traditional, organic and labelled products. However, as noted by some concerned actors, the original objective of taste education was the children's human development, not the promotion of particular products. The other side of this politicisation is 'scientification', in which taste education has been justified as an educationally effective approach. However, as a consequence, its original value has been reduced to certain measurable (often bio-psychological) indicators (improved sensibility, reduction of food neophobia, change in taste preference, etc.). This process was partly facilitated by its 'distinction' from traditional nutrition education by the taste education actors themselves, due to their particular emphasis on its 'non-normativity' and the use of 'sensory experiences'. Yet, as already challenged in Chapter 3, these pedagogical qualities have yet to be theoretically justified and, thus, need to be explored in further studies. Too much obsession with the above distinction could also obscure the other values of taste education, notably its sociological focus on the totality of food. The major implication is that, to overcome these three types of challenges (politicisation, scientification and distinction), it is necessary to go back to the basics of Puisais' philosophy, which is tackled in Chapter 8. Although taste education in Japan and Italy, as well as other concerned countries, has yet to reach this stage of institutionalisation, the French experience helps us to anticipate upcoming challenges.

Experiences of taste education can also have universal implications for food education. The experiences in three countries helped to highlight various effective strategies, from community dialogues to the empowerment of social structures (school meals and gardens, etc.), transdisciplinary practices in schools, the development of training programmes, partnerships with professional agri-food organisations and local authorities, and the establishment of evidence. Nevertheless, without the pedagogical (and even philosophical) attractiveness of taste education, it would have been impossible to put these strategies into effect. No other pedagogy has attained the equivalent level of appeal with taste education. Therefore, the studies on institutionalisation should not stop with institutional analysis but should rather go on to inquire into 'how food education should be', and its theoretical and philosophical foundation.

Notes

1 This chapter was reprinted with permission from Ueda, H. (2018b) A comparative analysis of promotion systems for taste education in Japan, France, and Italy. *Journal of Food System Research*, 25(2), 48–70.
2 Yet, technically speaking, the legal status of such responsibility has been contested because it serves as a 'quasi-imperative' by imposing some level of duty to undertake or refrain from certain actions, and justifies corresponding administrative directives and subsidisation.
3 Theoretically, taste education is value-neutral to any food (even sugar or artificial candy) and such criticisms have originated from misunderstandings that taste education is premised on certain norms of 'good food'. This issue is explored in detail in Chapter 8.

References

Billiaux, M. (2016) *Les Relais de l'Education au Goût en Languedoc-Roussillon-Midi-Pyrénées*. Université Toulouse Jean-Jaurès, Mémoire de 2ème année.

Borys, J.M., Le Bodo, Y., Jebb, S.A., Seidell, J.C., Summerbell, C., Richard, D., De Henaux, S., Moreno, L.A., Roman, M., Visscher, T.L.S., Raffin, S. & Swinburn, B. (2012) EPODE approach for childhood obesity prevention: Methods, progress and international development. *Obesity Reviews*, 13, 299–315.

Centro Educazione del Gusto & Nakano, M. (2012) *Mikaku no Gakko [School of Taste]*. Kirakusha.

Cody, E. (2009). Philosopher of taste. Article on November 7th on the website of Washington Post. https://www.washingtonpost.com/wp-dyn/content/article/2009/11/06/AR2009110604194.html (accessed in January 2019).

Corbeau, J.P. (1997) Socialité, sociabilité et sauce toujours. In J. Duvignaud, & Khaznadar, C. (eds.), *Cultures, Nourritures*, 69–81. Babel-Actes Sud.

Dupouy, P. (2012) *L'Education Sensorielle auprès des Enfants en Education Prioritaire: l'Example du Dispositif des Classes du Goût*. Université Toulouse Jean-Jaurès, Mémoire de 2ème année.

Dupuy, A. (2013) *Plaisirs Alimentaires, Socialisation des Enfants et des Adolescents.* Presses Universitaires de Rennes & Presses Universitaires François-Rabelais.

Durkheim, E. (1895) *Les Règles de la Méthode Sociologique.* Félix Alcan. Translated by T. Miyajima (1978) *The Rules of the Sociological Method.* Iwanami Shoten.

EPODE International Network (n.d.) http://epode-international-network.com/ (accessed in July 2017).

Fischler, C. (1990) *L'Homnivore.* Odile Jacob.

Freire, P. (1968) *Pedagogia del Oprimido.* Herder & Herder. Translated by M.B. Ramos (1996) *Pedagogy of the Oppressed.* Penguin Books.

Gaignaire, A. & Bourgeois, L. (2016) *BIOSENSO Programme.* Eveil'O'Goût.

Gaignaire, A. & Politzer, N. (2010) *Les Restos du Goût : 12 ateliers-repas au restaurant scolaire.* Yves Michel.

Gaignaire, A., Politzer, N., Monnery-Patris, S., Delaroche-Houot, S. & Dubois, D. (2011) *Evaluation des effets des Classes du Goût.* Ministère de l'Agriculture de l'Agroalimentaire et de la Forêt.

Girard, A. (2009) French cultural policy from André Malraux to Jack Lang: A tale of modernization. *International Journal of Cultural Policy*, 4(1), 107–125.

Glanz, K. & Rimer, B. (2005) *Theory at a Glance: A Guide for Health Promotion Practice (2nd edn).* National Cancer Institute.

Mauss, M. (1968) Les Techniques du Corps. In M. Mauss (ed.) *Sociologie et Anthropologie.* Presses Universitaires de France. Translated by T. Aji & Yamaguchi, T. (1976) *Sociology and Anthropology II*, 121–156. Kubundo.

McLeroy, K.R., Bibeau, D., Steckler, A. & Glanz, K. (1988) An ecological perspective on health promotion programs. *Health Education Quarterly*, 15(4), 351–377.

Ministère de l'Agriculture de l'Agroalimentaire et de la Forêt (2011) Education au Goût : Apprendre à goûter dés le plus jeune age. http://agriculture.gouv.fr/apprendre-gouter-des-le-plus-jeune-age (accessed in January 2019).

——— (2012) Education au Goût: Réseau national dédié. http://agriculture.gouv.fr/-education-au-gout-reseau-national-dedie (accessed in January 2019).

——— (2019) *Programme National pour l'Alimentation 2019–2023.* MAAF.

Ministère de l'Agriculture de l'Agroalimentaire et de la Forêt & Ministère de l'Education Nationale, de la Jeunesse et de la Vie Associative (2011) *Réferentiel de compétences: Educateur des Classes du Goût.* MAAF & MEN.

Ministère de l'Education Nationale, de la Jeunesse et de la Vie Associative (2020) Enseigner et éduquer l'importance du goût et la prise en compte des sens dans l'alimentation. https://eduscol.education.fr/cid148683/importance-gout-sens.html (accessed in June 2020).

Ministère des Solidarités et de la Santé (2019) *Programme National Nutrition Santé 2019–2023.*

Ministero Dell'Instruzione dell'Universita e della Ricerca (2015) *Linne Guida per l'Educazione Alimentare nella Scuola Italiana.*

Ministry of Agriculture, Forestry and Fisheries (2005) What is 'Shokuiku'? https://www.maff.go.jp/e/pdf/shokuiku.pdf (accessed in July 2017).

Minkler, M. & Wallerstein, N. (2012) Improving health through community organizing and community building. In M. Minkler (ed.) *Community Organizing and Community Building for Health (3rd edn)*, 37–58. Rutgers University Press.

Montoux, D., Martin-Poly, C. & Reverdy, C. (2012) *Programme de Formation des Enseignants sur les Classes du Goût.* Ministère de l'Agriculture de l'Agroalimentaire et de la Forêt & Ministère de l'Education Nationale, de la Jeunesse et de la Vie Associative.

Morin, E. (1973) *Le Paradigme Perdu : La nature humaine*. Seuil.

Niiyama, Y. & Ueda, H. (2020) France no senmon shokugyo (kan) soshiki to nogyo kyodo kumiai [Inter-professional organisations and agricultural cooperatives in France]. In Y. Niiyama (ed.) *Nogyo Keiei no Sonzoku, Shokuhin no Anzen [The Going Concern of Farm Management, Food Safety]*, 307–318. Showado.

Niiyama, Y., Takatorige, T., Sekine, K., Kawamura, R. & Kiyohara, A. (2014) Professional organisations in agri-food fields in France and the Netherlands. *Journal of Food System Research*, 20(4), 386–403.

Polony, N. (2010) Hommage aux maîtres. Article on January 3rd on the website of Figaro Blog Invité. https://blog.lefigaro.fr/education/2010/01/hommage-aux-maitres-suite.html (accessed in July 2019).

Politzer, N. (2004) Goûter à tout, cela s'apprend. Argos. http://www.educ-revues.fr/ARGOS/AffichageDocument.aspx?iddoc=32224&pos=0 (accessed in January 2019).

——— (2011) Où en est l'éducation au goût des jeunes en 2011. Dossier Interne. Translated by I. Tajiri. (2016) *Mikaku kyoiku no tanjyo to hensen* [The birth and transition of taste education]. In K. Ishii, Puisais, J., Sakai, N. & Tajiri, I. (eds.) *Kodomo no tameno Mikaku Kyoiku [Taste Education for Children]*, 2–13. Kodansha.

Poulain, J.P. (1985) *Anthropo-sociologie de la Cuisine et des Manières de Table*. Thèse pour le Doctorat de Sociologie, Universitaire de Paris VII.

——— (2002) *Manger Aujourd'hui, Attitudes, Norms et Pratiques*. Privat.

——— (2009) *Sociologie de l'Obésité*. Presses Universitaires de France.

——— (2017) *The Sociology of Food: Eating and the Place of Food in Society*. Bloomsbury.

Puisais, J. (1999) *Le Goût chez l'Enfant : l'Apprentissage en Famille*. Flammarion.

Reverdy, C. (2011) Sensory education: French perspectives. In V.R. Preedy, Watson, R.R. & Martin, C.R. (eds.) *Handbook of Behavior, Food and Nutrition*, 143–157. Springer.

Sasaki, T. (2008) *Slow Food Suginami Tokyo Sengen [The Declaration by Slow Food Suginami Tokyo]*. Siryusya.

Sato, M. (2009) Mikaku kyoiku wo toriireta chori gino shutoku no jyugyo dukuri [Development of the cooking class integrating Classe de Goût]. *Journal of Japan Association of Home Economics Education*, 51(4), 310–314.

Sato, M. & Ishii, K. (2014) An examination of cooking study of the French 'Sense of Taste Education'. *Journal of Japan Association of Home Economics Education*, 52(2), 85–93.

Schlich, P., Lange, C., Gaignaire, A., Reverdy, C. & Politzer, N. (2014) *Effet d'une éducation sensorielle sur les préférences et les comportements alimentaires d'enfants âgés de 8 à 10 ans*. Eveil'O'Goût.

Shiono, H. (2008) On basic acts. *Transactions of Japan Academy*, 63(1), 1–33.

Strozzi, M. (2015) Come Cambia l'Educazione Alimentare (1975–2015): Un'occasione mancata? In M. Montanari (ed.) *Cultura del Cibo* (vol. 4), 305–315. UTET Grandi Opere.

Takano, K. (2012) Business opportunity and food education in Japan. *Social Responsibility Journal*, 9(4), 516–533.

Togawa, R. (2009) France no shogakko ni okeru shokuiku [Food education in French elementary schools]. *Benesse Educational Research & Development Center*, 15, 40–45.

Tokumaru, S., Endo, Y., Aoki, M. & Okui, K. (2014) Practice of the taste educational program in junior high school home economics course. *Bulletin of Joetsu University of Education*, 33, 205–216.

Ueda, H. (2017) Impact evaluation of a school-based food education intervention on cognitive and attitudinal aspects of children's dietary habits: A case study of a sensory education program 'Leçon de Goût'. *Journal of Food System Research*, 24(1), 2–17.
——— (2018a) Impact evaluation of the updated educational programme of taste education with its evaluation system improved based on the development of 'food curiosity' scale. *Journal of Japanese Society of Shokuiku*, 12(2), 109–124.
——— (2018b) A comparative analysis of promotion systems for taste education in Japan, France, and Italy. *Journal of Food System Research*, 25(2), 48–70.
Ueda, Y., Otahabara, M. & Nagai, A. (2015) Implementation of palate education on fourth graders and its effect evaluation. *Journal of Japanese Society of Shokuiku*, 9(3), 247–255.
Zimmerman, M. (2000) Empowerment theory: Psychological, organizational, and community levels of analysis. In J. Rappaport & Seidman, E. (eds.) *Handbook of Community Psychology*, 43–59. Kluwer Academic/Plenum.

Part II

Gastronomy as Taste Education's Philosophical Tradition

6 History of French Gastronomy

In Part I, I presented the difficult challenge of articulating the connection between taste education and gastronomy. As an entry point into such complexity, this chapter comprises an overview of the history of the genesis and transmission of French gastronomy by mobilising a rich body of knowledge in sociology, history and anthropology. This overview ultimately reveals that the content and social position of contemporary gastronomy is becoming increasingly complicated. To demonstrate such complexity, I continue by analysing the recent heritisation of gastronomy by UNESCO by comparing it with that of *washoku* in Japan. I begin this historical analysis mainly with Poulain's (2017) comprehensive illustration of its philosophical, social, cultural and religious conditions, in addition to other relevant literature.

6.1 Genesis: The Civilising Process at Versailles

6.1.1 *Dialectic Dynamism in the Intellectual Realm*

Thinkers since antiquity until the Renaissance eloquently pondered the art of eating well, as we can see in examples from Plato's *Symposium*, Athenaeus' *Deipnosophistae* and Rabelais' *Gargantua*. However, in the 17th century, the practice of eating was given an ambiguous status due to a double exclusion: (1) The Cartesian distinction between subject and object, which is one of the foundations of modern science, led to the indirect consequence of putting the experience of eating (at the intersection between mind and body) outside the philosophical realm; (2) Its strong corporeality prevented the aestheticisation of eating from occupying a legitimate space in the artistic domain. This marginalisation of gastronomic discourse undoubtedly characterised its unique deployment (Poulain 2017, pp.184–186).

A new reorganisation of this academic map had to wait for two periods: In his book *The Physiology of Taste* (1825), Brillat-Savarin theorised food and taste as non-corporeal and intellectual subjects, which partially contributed to the legitimisation of gastronomy in academic culture. However, it was not until the 1960s that gastronomy finally became a legitimate scientific subject. This period was marked by the seminal works of Barthes (1961) and Lévi-Strauss (1965), and

DOI: 10.4324/9781003341895-9

subsequently by two 'discoveries'. The first was the French translation of the work of Norbert Elias (originally published in 1939) and Barthes' intellectual revisitation of gastronomic critics (1975). This academic interest was taken up by the historians of Annales (Aron 1973; Flandrin & Montanari 1996), the sociologists of food (Mennell 1985; Poulain 1985, 2002; Fischler 1990) and geographers of taste (Pitte 1991), and was followed by an immense increase in gastronomic studies. Helped by this interdisciplinary interest, the 'totality' of gastronomy as being at the intersection of bio-psycho-socio-cultural entities was finally recognised in the academic world.[1]

To recall its intellectual history, I claim that the Cartesian division was not sterile at all. Inversely, it played an indispensable role in driving the dialectical dynamism forward, whereby the status of the connection between food and the mind shifted from being considered an undivided one (Antiquity to Renaissance) to completely divided (Cartesian) and then as a total social fact (1960s and onwards). In contrast to French gastronomy, as we shall see later, such dialectics were absent from the development of Japanese gastronomy.

6.1.2 Social Distinction and the Sophistication of Table Manners

In his book *The Civilising Process* (1969), Elias saw the appearance of the concept of 'civility' as the backbone of court society. This civility encompassed a set of codes of behaviour that ensured that the manners of the members of the upper class reflected their social status. One of the most important categories of etiquette was table manners; at meal times, the body was on display and these codes defined the conditions in which food was incorporated. In France, the expression of civility acquired a unique character. The establishment of the Court at Versailles in the late 17th century instigated an effective civilising process. The Court continued the process of centralising the state and attracted the provincial nobility. As a consequence of the subsequent local power vacuum, the increasingly affluent bourgeoisie determinedly copied the manners of the aristocracy – and the latter refined their manners further to distinguish themselves from the former. This game of social distinction between the upper and the lower social classes led to highly sophisticated art of the table (Poulain 2017, pp.186–187).

6.1.3 The Alchemical Approach and Sophistication of Taste

In the late 16th century, following the development in navigation and the discovery of the so-called 'New World', spices came to be greatly popularised and ceased to be a marker of social distinction. Having abandoned the use of spices as a sign of luxury, French gastronomy was henceforth to demonstrate its sophistication through an emphasis on the tastes of the food items themselves. In *Les délices de la campagne* (1654), Nicolas de Bonnefons presented a revolutionary principle: 'A cabbage soup should smell of cabbage [...] What I say of soups I mean to be understood as applicable in every instance and to serve as a general rule for everything that is eaten'. There was a major shift from the medieval culinary style based on masking to a cuisine based on creating associations and harmonies of taste. In this way, a new culinary category now emerged, the purpose of which was to enhance the flavours of food with stocks and sauces.

Behind this quest for taste, the alchemical approach, in which the relationship between man and the universe was viewed as a symbolic correspondence, served as a theoretical model for the 18th century chefs (even the 19th century chefs, together with the emerging experimental sciences). Culinary books of that period were characterised by genuinely alchemical language to express the aim of improving stocks and sauces (Neirink & Poulain 1988). Yet, alchemical research was not limited to the quest for the most perfect sauce; in improving their cooking, chefs believed that they were also improving themselves and, moreover, that they were contributing to the progress of humanity. It was due to this alchemical quest for good taste that French cuisine acquired greater complexity (Poulain 2017, pp.188–190).

6.1.4 Catholicism and Justification of Pleasure of Eating

In counterpoint to Max Weber's *The Protestant Ethics and the Spirit of Capitalism*, Poulain (2017) argues that the aestheticisation of eating owes much to the moral stance of Catholicism. In Christianity, there are three opposing attitudes to food: asceticism, moderation and hedonism. Poulain asserts that Catholicism is characterised by hedonism, in which the enjoyment of earthly pleasures is viewed as a means of glorifying the works of God.

There are three points which constitute the spiritual infrastructure that organises the relationship between Catholicism and the pleasure of eating. The first transformation occurred in the ritual of Holy Communion. It used to be celebrated using the two elements of bread and wine, which were believed to become Christ's Body and Blood. Traditionally, both the clergy and the laity partook of both elements. However, from close to the end of the Middle Ages, only the former was permitted to consume both the bread and wine. The second major change occurred as the Reformation took hold. It involved abandoning the use of leavened bread for the host, and replacing red wine with white (although not uniformly in France), which established a symbolic distance between Christ and food. Moreover, with the discovery of the 'New World' and of the cannibalism practiced by some of its habitants, the conflict over the theories of Christ's presence in the bread and wine was intensified. As a reaction to such criticism, the ritual was altered and a division between the secular consumption of food and its use in sacred acts was established. In this way, the daily act of eating came to be disconnected from the authority of the Church. Yet, gluttony still remained one of the seven deadly sins. Therefore, as the third transformation, the elaboration of eating practices was expected to introduce the distinction between gluttony, governed by the senses, and gastronomy, which was seen to be ruled by the spirit (pp.190–195).

6.2 Transmission: From Post-Revolutionary to Post-War Period

6.2.1 *The Emergence of Restaurants and the Social Organisation of Chefs*

All the above conditions led to the genesis of an appreciative attitude to the totality of food and the pleasure of eating, as well as sophisticated table manners, taste and culinary systems, all of which became the indispensable constituents of French gastronomy. During the *Ancien Régime*, gastronomy was still an aristocratic privilege; however, after the French revolution, it came to be appreciated by other social classes as well. One of the major drivers for such descending transmission was the emergence of restaurants opened by chefs who had previously served the aristocracy (Aron 1973; Pitte 2002). The surge of restaurants was noted with surprise by Grimod de la Reynière: 'One could not count a hundred of them before 1789. Today, there are perhaps five times as many' (1803). However, the function of restaurants also changed. While restaurants during the *Ancien Régime* used to serve set meals or soups for recuperation, the post-revolutionary ones, as meticulously observed by Aron (1973), took on a new function, namely that of creating the space for social distinction by the emerging bourgeoisie.

It was not only restaurants, but also tables at the bourgeoisie's homes that became a field of social distinction. One of the representative figures who played such a role in gastronomy *à la maison* was Antoin Carême (1783–1833). While serving numerous royal and bourgeois families, he pursued his passion for writing and produced three books: *Maitre d'hotel Français*, *Le Patissier Royal Parisien* and *L'Art de la cuisine française au XIXe siècle*. These books reached out to cooking professionals and the bourgeoisie of his period, who looked for competitive domestic chefs as a sign of social status.

Nevertheless, as was also the case with Carême, most 19th century chefs were from poor backgrounds and their working conditions were extremely harsh. They faced three major challenges: the establishment of a system of professional education, the improvement of working conditions and the social recognition of chefs as professionals (not domestic workers). Auguste Escoffier (1846–1935) was one of the important figures who contributed enormously to improving the social status of chefs, both technically and symbolically: He formalised classical French cuisine, including techniques and restaurant management, and published *Le Guide Culinaire* in 1901, which was to serve as the major reference manual until the Nouvelle Cuisine paradigm emerged. It was not until the 1920s that the professional status and social institutions of chefs were finally established (Drouard 2004).

6.2.2 *Gastronomic Critics and the Legitimisation of Taste*

The first appearance of the word 'gastronomy' is attributed to Joseph de Berchoux's poetry in 1801. He might have been attracted by Archestratos' *Gastronomia*, a poem that contained a catalogic knowledge of food, taste and cooking techniques of every region in Greece during the 4th century BC. In the 19th century, a new category of gastronomic guides and literature abounded, which was initiated by gastronomic critics such as Grimod de la Reynière and Brillat-Savarin.

As a member of the highest aristocratic circles but excluded from Paris due to his misconduct, Grimod returned to the post-revolutionary capital and took stock of new social phenomena around the table. This led him to publish *L'Almanach des gourmands* (1803) and *Manuel des amphitryons* (1808). The former was intended to 'guide the middle classes through the labyrinth of the new food shops, restaurants and catering enterprises', whereas the latter was addressed to those 'newly enriched by the Revolution [...] by causing these riches to change hands, had placed them at the disposal of men who, until now, had been strangers to the art of using them, and of enjoying them in a noble fashion'.

On the other hand, Brillat-Savarin, whose origin was far more modest, played numerous roles, such as attorney, politician, sometimes musician and ideologue. It is particularly notable that, prior to the well-known publication of *The Physiology of Taste*, he was concerned mainly with politics and even wrote *Vues et projets d'économie politique* in 1801, in which he proposed the establishment of professional educational institutions for future politicians and administrators. It is in this political and educational context that his philosophy of taste education should be examined (which is done in Chapter 8).

Thanks to the dual nature of their social positions, these gastronomic critics played the important role of decoding aristocratic practice so that the bourgeoisie could copy it. This system of legitimisation of 'good taste' which they established was to serve as an important criterion in contemporary food guides, reviews and gastronomic literature (Poulain 2017, pp.195–197).

6.2.3 *Tourism Development and Regional Gastronomies*

I have so far described the 'descending' transmission of French gastronomy from the upper to lower social classes. However, the process of transmission is not so simple because a sort of 'ascending' effect can also be observed. During the *Ancien Régime*, gastronomy was characterised by its distance from necessity and want, and was founded on the repression of regional and popular cuisines, which were excluded from such high gastronomy.

This opposition was increasingly disturbed as two social movements gradually modified the social representation of gastronomy itself (Csergo 1996; Poulain 2005, 2017, pp.11–16): The first, the regionalist movement, emerged as a counter-reaction to the dissolution of the provinces during the Revolution, under which food traditions, though their emblematic nature, became the site of cultural resistance. This regionalist climate continued and was well orchestrated towards the latter half of 19th century. Numerous local publications about regional dietary traditions, which catered for local readers, were disseminated. At the beginning of 20th century, the capital followed this movement by issuing volumes with a similar purpose of celebrating regional culinary diversity in France.

This appetite for regional cuisine was further boosted by the development of tourism from the 1920s. Although the popularisation of tourism, whereby ordinary people travelled extensively, did not occur until the post-war period, opulent singles and professional men (e.g., doctors and attorneys) started to enjoy driving to regions and appreciating their unique recipes. Curnonsky (1872–1956) was lucky enough to be in such a social position. As a single journalist, he drove around to explore regional recipes and published *Tour de France gastronomique* (28 volumes in total) and *Le trésor gastronomique de France*. The co-author of the latter book, Austin de Croze (1866–1937), is another notable figure in this development. He organised a regional gastronomy week (1923–1924), at which chefs from all over France were invited to present their regional dishes. This period was also marked by a growing number of gastronomic associations created by regional or Parisian elites (such as *Académie des Gastronomes* and *Le Club des cent*). These social movements eventually modified the previous representation of gastronomy and facilitated the successful integration of regional and popular cuisine, which had been previously banished, into the gastronomic realm. This is despite the fact that this development seemed to contradict the intensifying conservatism in the kitchen started by Escoffier during the same period.

6.2.4 *The Expanding Middle Class and Nouvelle Cuisine in the Post-War Period*

From the late 1960s, the Nouvelle Cuisine movement played a pivotal role in further modifying the representation of gastronomy and transmitting it to a much wider population. This movement was shaped by emblematic chefs such as Paul Bocuse, Michel Guérard and the Troisgros brothers, who denounced the chefs' lack of autonomy, which hindered their inventiveness, in Escoffier's classical cuisine, while appropriating his idea of simplicity and economy, to which he applied a new logic. As a source of invention, these evangelist chefs were eager to incorporate not only regional or popular French cuisine into their offerings, but also dishes from around the world (notably Japan) and the former colonies and immigrants (Rao et al. 2003).

Two influences can be identified to explain the emergence of the Nouvelle Cuisine (Aron 1997): The first was the expansion of the middle class due to post-war economic development. The members of this class became the major recipients of this new culinary paradigm. The second was the protests of May 1968 against elite cultures, following which it became impossible to publicly defend high gastronomy. Under these social movements, French gastronomy came to be characterised by a return to nature (as opposed to culture), an ascetic (as opposed to hedonistic) attitude to food and the increasing dependency on guides (rather than one's own tastes). One can pessimistically view this change as the 'glacialisation' of gastronomy (Aron 1997).

Nevertheless, there is no doubt that the Nouvelle Cuisine movement accelerated the 'gastronomisation of regional cuisines' (Poulain 2012), which was further advanced by the interdisciplinary study of gastronomy from the 1980s, the

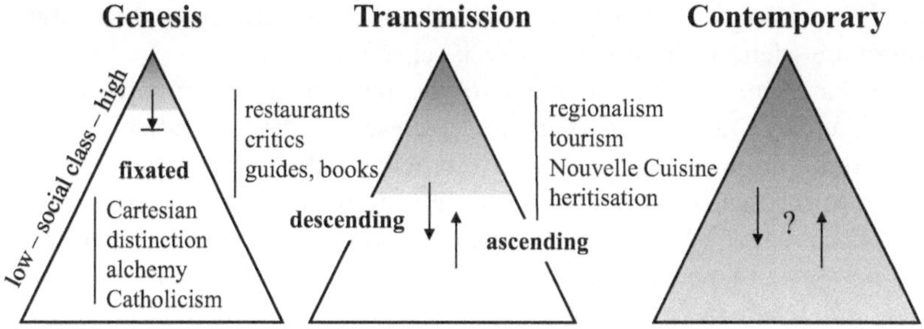

Figure 6.1 Transmission of French Gastronomy
Source: Developed by the author

creation of an inventory for regional recipes by CNAC in the 1990s and the food heritisation by UNESCO in 2010 (for a visual summary, see Figure 6.1).

6.3 Heritisation and Challenges in Defining Food Cultures

Although previous gastronomisation or heritisation projects were organised more or less on a regional scale, UNESCO's recent project took place on a national scale; thus, it caused a new social and scientific controversy worthy of closer analysis. In France, gastronomy ('gastronomic meal of the French': GMF) was inscribed as an intangible cultural heritage in 2010, followed by numerous food heritages, including Japan's *washoku* ('traditional dietary cultures of the Japanese; notable for the celebration of New Year') in 2013. Hereafter, based on its detailed analysis (Ueda & Niiyama 2019), I outline the process and results of defining these food cultures, with the purpose of articulating challenges inherent in contemporary gastronomies, both in France and in Japan.

6.3.1 *Problems with the Definition Process*

Confusion was evident in both countries during the process of determining definitions, perhaps because the rationale for UNESCO's intangible cultural heritage was not fully understood. In France, the project initiators initially intended to inscribe high gastronomy, partly under the influence of a political intention to advocate its cultural excellence. However, responding to UNESCO's request for a more popular social practice, the working group shifted the focus of inscription from high gastronomy to the GMF, by which its inherent elitism was deemed to be eliminated (Tornatore 2012). The inadequacy of the definition of the GMF was highlighted by the working group members, who had to eliminate descriptions about products and producers from the definition to adhere to the UNESCO guidelines; any economic attempt to safeguard products or producers could not be justified (Csergo 2016). Moreover, Naulin (2012) argues that this definition of the GMF was an 'artefact of the compromise' between the project initiators' vision of high gastronomy and UNESCO's popular conception of heritage, and the result of neither pure scientific discussion nor social agreement.

Much debate can also be identified in the Japanese case. Initially, *kaiseki* (Japanese haute cuisine) was proposed by a Japanese culinary association in Kyoto for the working group to inscribe as an intangible cultural heritage. However, responding to the fact that South Korea's application to register the 'royal court cuisine' was halted by UNESCO, the Japanese working group decided to shift its focus from *kaiseki* to a more popular and everyday food culture, *washoku*. However, Cwiertka and Yasuhara (2016) point out that this makeshift definition resulted in 'modifying historical facts'. Indeed, certain components, namely the everyday consumption of rice at a national level, a particular meal structure, namely *ichi-ju-sansai* (rice, soup and three dishes), and beautiful decoration with equipment or flowers, were normalised as 'traditional' in the definition of *washoku*. Yet, in reality, it was not until the post-war economic development that the ordinary Japanese had managed to eat in this manner. There were other criticisms as well: Fujihara (2016) pointed out that the components of this definition were not exclusive to *washoku* and could be true of other food cultures. Moreover, its inter-relational development in tandem with other countries was neglected in the definition. Yasui (2018), compiling claims made in previous studies, warned us about the gendered, patriarchal and ethnically exclusivist ideologies underlined in the state-run programme to institutionalise *washoku*.

As noted above, previous studies helped to clarify several problems with the process of defining these food heritages. However, no matter how controversial it was, two implications can be identified from this social project: First, it is one of the rare projects in which concerned experts were dedicated to 'defining food cultures' at a national level. Thus, if its political and ideological effects are carefully considered, it serves as a valuable source for understanding these contemporary food cultures. Second, more practically speaking, it is the final product (definition) that bears institutional importance, to be protected nationally, on the one hand, and to be promoted internationally, on the other. Therefore, no effective safeguarding measure can be achieved without scrutinising these final definitions.

6.3.2 *Conceptual Structures of Gastronomy and Washoku*

The nomination files of UNESCO (2010, 2013) were analysed using a content analysis method to identify the definitions of gastronomy and *washoku*. This method is a systematic qualitative approach to codifying each word or phrase, categorising those with similar meanings and identifying the interrelationship of these categories (constituent concepts). Eight constituent concepts were identified by means of this analysis: the seven common categories of 'social practice', 'social integration', 'identity', 'diversity', the 'human-nature relationship', 'tradition' and '(extra)ordinariness of occasions', and the eighth category of 'pleasure' for gastronomy and 'health' for *washoku*. Note that the commonality of the categories does not necessarily mean that they have common meanings. These conceptual structures are illustrated in Figure 6.2. Here, I summarise the contrasting features of gastronomy and *washoku* in relation to the resultant challenges related to their

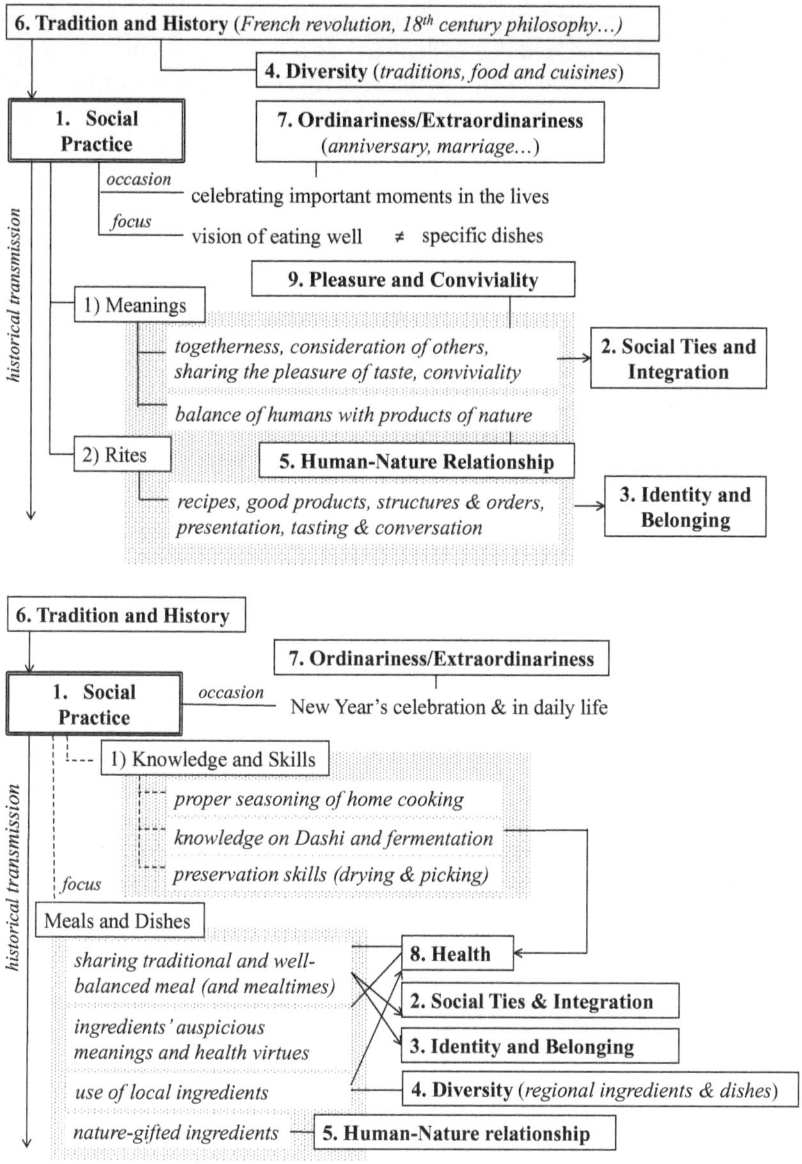

Figure 6.2 Conceptual Structures of Gastronomy and Washoku

Source: Reprinted with permission from Ueda and Niiyama (2019, p.151) with a partial revision by the first author. The upper figure refers to gastronomy, whereas the bottom one concerns *washoku*.

safeguarding measures (for the details of each constituting concept, see Ueda & Niiyama 2019).

First, 'social practice' was the most pivotal of the eight constituent concepts, and gastronomy was defined as a social practice characterised by specific meanings and rites that were then systematically concretised. In contrast, the social practice of *washoku* was too unclear and exemplified with only a few ingredients and dishes (*dashi* stock, pickled vegetables, etc.), as if it were a 'tangible' heritage. This confusion was repeated in the Japanese safeguarding measure, the aim of which was to (re)define the protection area of *washoku*. However, the task force for safeguarding was locked into a discussion about whether certain hybridised dishes (such as *ramen*) should be considered *washoku*.

Second, in terms of occasions of such practice, gastronomy was defined as 'festive' meals for celebrating life events (anniversaries, marriage etc.) and it was further characterised by other extraordinary features (e.g., the use of unusual recipes). On the contrary, the occasions of practising *washoku* were ambiguous; it was explained as 'a part of daily life', while it was contradictorily described with extraordinary examples, notably the celebration meal for New Year. This ambiguity must have been unexpected for the Japanese working group, which initially aimed to inscribe ordinary practices of daily life.

Third, a traditional character was indispensable for UNESCO's cultural heritages, and the history of gastronomy was specified as having undergone a transcending transmission from higher to lower society across generational, class and philosophical differences, through the French revolution and 18th century philosophy. This historical understanding is mostly consistent with what was outlined in this chapter, although the conclusion about the degree of its transcendency might be questioned. Surprisingly, no specific historical period was described for *washoku*, and this resulted in logical contradiction, particularly regarding its 'healthy' feature: From the perspective of nutritional science, the meal structure with an ideal PFC (protein, fat and carbohydrate) balance refers only to that of the 1980s, not to meals before this period (F-deficiency and P-overabundance) or to those after this (F-overabundance) (MAFF 2009). In terms of this logic, the meals of only 40 years before the inscription were considered 'traditional', which resulted in the unintended consequence of 'modifying historical facts' (Cwiertka & Yasuhara 2016). The same contradiction is also evident in the 'view of nature' as an essential spirit of *washoku*, because the influences of mores, religions and philosophies (Shintoism, Buddhism, Confucianism etc.) and classes (court aristocracy, warriors, merchants, etc.) were all absent from the definition. In Chapter 9, I argue that this is not a problem of description only in the nomination file, but rather one that is inherent in Japanese food culture studies.

Lastly, the process produced an unexpected institutional problem unique to *washoku*, which is the discrepancy between the semi-final definition that was proposed by the Japanese working group and UNESCO's final definition. Indeed, the former definition consisted of four elements: (1) various fresh ingredients and the use of their natural tastes; (2) well-balanced and healthy diets; (3) an emphasis on the beauty of nature and the changing of seasons in the presentation of dishes; and (4) connecting to annual events based on the essential spirit of respect for nature. The final definition (consisting of eight constituting concepts) turned out to be clearly different from the latter one. The concern was that the normalisation of the former four-layered definition in the safeguarding measure would deviate *washoku* from the internationally-authorised definition.

6.4 Concluding Remarks

Today, gastronomy is often considered an elitist practice and, if taste education is the daughter of French gastronomy, this pedagogy has to face such scepticism. As was outlined in this chapter, French gastronomy was first birthed in the Court at Versailles and later transmitted to the wider population through multiple media, such as the post-revolutionary restaurants, gastronomic critiques and literature, and tourism, and finally became the acclaimed 'common property' of the French. Behind this, the dynamism of social distinction between social classes that characterised its genesis has continued to be active in driving the transmission of gastronomy.

It is partly due to this complex historical and social mechanism that gastronomy is still considered an elitist practice. However, this criticism might also be due to the debate at a more theoretical level: Bourdieu (1979) was one of those theorists who asserted the autonomy of working-class taste for 'necessity' that was distinguished from the elite taste for 'form' in gastronomic practice. In contrast to Bourdieu and his followers (Grignon & Grignon 1980), Poulain (2017, pp.187–188) points to the historical fact of the descending transmission of gastronomy. From a critical but more eclectic perspective, Corbeau (1991, p.12) claims that gastronomy can be conceived purely as the 'ritualisation' of a sensory emotion that exists in universal social dimensions. My interpretation is that all of these theorists posit an aspect of the truth: On the one hand, gastronomy has descended from the elite to the whole social body but, on the other hand, it took different or common forms according to social status. To resolve this enigma, it is essential to investigate what gastronomy means to the French population in contemporary society. This question was half-raised during UNESCO's heritisation project, yet it still requires a more scientific exploration. At the same time, the history of French gastronomy has an important implication for Japanese *washoku*. In Chapter 9, I adopt a comparative approach to depict the philosophical, social and religious histories of Japanese gastronomy, while addressing unanswered questions regarding its healthiness and unique view on nature that were problematised in this chapter.

Note

1 The thematisation of gastronomy in the social sciences has a close link with that of food in general (Poulain & Corbeau 2012).

References

Aron, J.P. (1973) *Le Mangeur du XIXe Siècle*. Robert Laffont.

——— (1997) De la glaciation dans la culture en général et dans la cuisine en particulier. In J. Duvignaud & Khaznadar, C. (eds.) *Cultures, Nourritures, Internationale de l'Imaginaire*, 13–37. Babel/Actes Sud.

Barthes, R. (1961) Pour une psycho-sociologie de l'alimentation contemporaine. *Annales*, 16(5), 977–986.

——— (1975) Lecture de Brillat-savarin. In J.A. Brillat-Savarin (ed.). *Physiologie du Goût*, 7–33. Hermann. Translated by R. Howard (1989) *The Rustle of Language*, 250-270. University of California Press.

Bourdieu, P. (1979) *La Distinction: Une Critique Sociale du Judement*. Editions de Minuit.

Brillat-Savarin, J.A. (1825) Physiologie du goût. Edited by J.F. Revel (1982) Flammarion.

Corbeau, J.P. (1991) *Essai de Reconstruction Utopique des Formes et dans Jeux du Manger*. Thèse de Doctorat. Université de Paris VII.

Csergo, J. (1996) L'Emergence des Cuisines Régionals: France. In J.L. Flandrin & Montanari, M. (eds.) *Histoire de l'alimentation III*. Fayard. Translated by M. Kitadai (2006) The emergence of regional cuisines: France. In N. Miyahara & Kitadai, M. (eds.) *History of Food III*, 1080–1098. Fujiwara Shoten.

——— (2016) *La Gastronomie est-elle une Marchandise Culturelle comme une Autre? La Gastronomie Française à l'UNESCO : Histoire et Enjeux*. Menu Fretin.

Cwiertka, K. & Yasuhara, M. (2016) *Himerareta Washokushi [The Hidden History of Washoku]*. Sinsensha.

Drouard, A. (2004) *Histoire des Cuisiniers en France : XIXe – XXe siècle*. CNRS Editions.

Elias, N. (1939) Über den Prozeß der Zivilisation: soziogenetische und psychogenetische Untersuchungen. Haus zum Falken. Translated by E. Jophcott (2000) *The Civilizing Process: Sociogenetic and Psychogenetic Investigations (2nd ed.)*. Blackwell Publishing

Fischler, C. (1990) *L'Homnivore*. Odile Jacob.

Flandrin, J.L. & Montanari, M. (1996) *Histoire de l'Alimentation*. Fayard.

Fujihara, T. (2016) *Shoku wa kyoiku no kadai nanoka* [Is food a matter of education?] In T. Sato (ed.) *Gakushu suru Shakai no Asu [The Future of Learning Society]*, 179–202. Iwanami Shoten.

Grignon, C. & Grignon, Ch. (1980) Styles d'alimentation et goûts populaires. *Revue Français de Sociologie*, XXI, 531–569.

Grimod de la Reynière, A.B.L. (1803) *L'Almanach des Gourmands*. In J.L. Bonnet (ed.) *L'Almanach des Gourmands: Texte Intégral des Huit Années 1803–1812* (2012). Menu Fretin.

Lévi-Strauss, C. (1965) Le Triangle Culinaire. *L'Arc*, 26, 19–29. Translated by P. Brooks (1966) The Culinary triangle. *Partisan Review*, 33, 586–595.

Mennell, S. (1985). *All Manners of Food: Eating and Taste in England and France from the Middle Ages to the Present*. Basil Blackwell.

Ministry of Agriculture, Forestry and Fisheries (2009) *Annual Report on Food, Agriculture and Rural Areas in Japan 2009*. MAFF.

Naulin, S. (2012) Le repas gastronomique des Français : Génèse d'un nouvel objet culturel. *Sciences de la Société*, 87, 8–25.

Neirink, E. & Poulain, J.P. (1988) *Histoire de la Cuisine et des Cuisiniers* (2nd ed.). Editions Jacques Lonore.

Pitte, J.R. (2002) *French Gastronomy: The History and Geography of a Passion*. Columbia University Press.

Poulain, J.P. (1985) *Anthropo-sociologie de la Cuisine et des Manières de Table*. Thèse pour le Doctorat de Sociologie, Universitaire de Paris VII.

——— (2005) French gastronomie, French gastronomies. In D. Goldstein & Merkele, K. (eds.) *Culinary Cultures of Europe Identity, Diversity, and Dialogue*, 157–170. Editions du Conseil de Europe.

——— (2011) La gastronomisation des cuisines de terroir: sociologie d'un retournement de perspective. In N. Adell & Pourcher, Y. (eds.) *Transmettre, Quel(s) Patrimoine(s)? Autour du Patrimoine Culturel Immatériel*, 239–248. Michel Houdiard.

——— (2012) Gastronomisation des cuisines de terroir. In J.P. Poulain (ed.) *Dictionnaire des Cultures Alimentaires*, 622–627. Presses Universitaires de France.

——— (2017) *The Sociology of Food: Eating and the Place of Food in Society*. Bloomsbury.

Poulain, J.P. & Corbeau, J.P. (2012) Thèmatisation de l'alimentation dans les sciences humaines et sociales. In J.P. Poulain (ed.) *Dictionnaire des Cultures Alimentaires*, 1329–1336. Presses Universitaires de France.

Rao, H., Monin, P. & Durand, R. (2003) Institutional change in Toque Ville: Nouvelle Cuisine as an identity movement in French gastronomy. *American Journal of Sociology*, 108(4), 795–843.

Tornatore, J.L. (2012) Retour d'anthropologie: le repas gastronomique des Français: Eléments d'ethnographie d'une distinction patrimoniale. *Ethnographiques.org*, 24, 1–29.

Ueda, H. & Niiyama, Y. (2019) Articulating challenges in defining Japanese washoku and French gastronomy: Comparative analysis of inscribed definitions and their safeguarding measures. *Journal of Food System Research*, 26(3), 144–164.

United Nations Educational, Scientific and Cultural Organization (2010) *Gastronomic Meal of the French* (nomination file, No. 00437).

——— (2013) *Washoku; Traditional Dietary Cultures of the Japanese, Notably for the Celebration of New Year* (nomination file, No. 00869).

Yasui, D. (2018) Shokubunka no kata: Bunkaisan toshiteno washoku [The pattern of food cultures: washoku as cultural heritage]. In M. Akitsu, Sato, Y. & Takenouchi, H. (eds.) *Shoku to No no Atarashii Rinri [A New Food and Agricultural Ethics]*, 145–169. Showado.

7 Contemporary French Gastronomy

For UNESCO, gastronomy is defined as a 'social practice of eating well at festive occasion'. However, a closer look reveals profound differences. We are far from a consensual definition. In line with UNESCO, some people, such as the creators of the slow food movement, claim that gastronomy is for everyone. They see it as an 'art of living' and assert its vocational universality. In contradiction with this popular vision, others emphasise the know-how and tasting skills. For them, gastronomy is a matter of experts, such as great restaurateurs and their suppliers, and connoisseurs-consumers.

In order to research the contours of the gastronomic phenomenon, several legitimate entry points are possible: professionals (producers, craftsmen in the food industry, chefs, maitres d'hôtels, sommeliers), gastronomic critics, observers of society (such as historians, sociologists and anthropologists), and finally the public. All of them carry a part of the truth. The choice made in this chapter is to explore the representations of the French public.[1] I first justify this analytical focus.

7.1 Missing Dimension: The French Public

7.1.1 *Who Produces Gastronomic Discourses?*

In the preceding chapter, I discussed how the polysemicity of gastronomy has been produced in its complex history of genesis and transmission and the studies of various disciplines since the 1960s. Nevertheless, when a focus is on 'who' produces such gastronomic discourses, at least three different dimensions can be identified.

First, the *normative* dimension refers to the gastronomic discourse that decides what is good food and taste. It was initiated by the gastronomic critics, as introduced in Chapter 6, such as Grimod and Brillat-Savarin who knew the two worlds (aristocracy and bourgeois), and repeated until today in the form of gastronomic guides, magazines and culinary books.

Second, the *academic* dimension does not ostensibly make such normative judgement but, rather, analytically decodes and creates the gastronomic discourse. It is particularly so in our 'thinking society' (Moscovici 2001) where scientific knowledge is to be gradually transformed into common knowledge.

DOI: 10.4324/9781003341895-10

However, careful attention should be paid to the epistemological status of the evidence used for producing such academic discourse. Indeed, the previous academic definitions of gastronomy founded primarily on the normative discourse (gastronomic guides and culinary books), not the *public* discourse, despite the prevailing notion that gastronomy has become common property for the entire French population.

This gap of knowledge around the *public* dimension of gastronomy recently caused a social controversy when inscribing it as UNESCO's (2010) intangible cultural heritage. In this project, the public dimension of gastronomy was partly captured through pilot interviews (Csergo 2016, pp.191–192) and the follow-up quantitative survey with the French (n = 998) with some closed questions on 'gastronomic meal of the French' (Mathé et al. 2009). These studies provide us with useful insights for its lay definition, but the socio-political nature of this project has been criticised as having inhibited the pure scientific formulation of such definition. Therefore, in this chapter, I aim to detach, for a while, from the context of recent heritisation with UNESCO and produce a more scientifically reliable public definition of gastronomy in the French population.

7.1.2 Survey Design

To elucidate the contemporary definition of gastronomy, two kinds of surveys were conducted. First, exploratory semi-structured interviews with French citizens were conducted during the period from January to December 2018. The interviewees consisted of three different social groups: Actors involved in the *Cité de la Gastronomie* (Gastronomic City) project, a national safeguarding measure for gastronomic heritages (Tours: n = 3, Paris: n = 3, Dijon: n = 3); professionals in the agri-food sector (n = 10 in Toulouse such as a farmer, butcher and bread baker); and non-professionals (n = 30 in Paris, Tours, Toulouse and Lyon). The interview for each group was conducted respectively for 60–90 mins, 30 mins and 1–5 mins, with the following protocol: (1) Explaining the interview objective and protocol; after obtaining their informed consent, (2) posing the first question, 'what does gastronomy mean to you ...?'; and (3) asking subsequent questions (such as when, where, with whom and what recipe) to concretise the aspects of eating model in their definition of gastronomy. Although the detailed results of these preliminary interviews are not included in here, some relevant responses were selectively referenced to better inform the second results described below.

Second, an online questionnaire was completed by the French population (aged 18–74). The questionnaire was developed based on the exploratory interview and methodological perspectives from the sociology of food (detailed later). The panel was purchased from a private research company, and the questionnaire was sent to this panel by email in March 2019. The respondents, who showed an interest in this survey and agreed with its purpose, completed the questionnaire sheet for averagely 15 mins, and 438 effective responses were finally collected. Their demographic characteristics largely represented the national population (Table 7.1).

Table 7.1 Sample Characteristics in Comparison to the French National Population

Gender	Female	Male		
Sample	51.6	48.4		
National population	51.6	48.4		
Age	Young (18–34)	Middle (35–54)	Old (55–74)	
Sample	30.1	42.7	34.9	
National population	32.3	45.6	32.1	
Profession	Farmer, craftsman, shopkeepers	Executive, higher intellectual	Intermediate profession	Clerk
Sample	4.6	15.1	12.6	28.1
National population	4.2	9.6	13.8	15.3
	Manual worker	Retired	Student	Unemployed
	3.7	17.6	5.9	12.6
	11.8	32.5	7.8	4.8
Household income	High	Middle	Low	
Sample	17.1	30.8	52.1	
National population	20	50	30	
Education	Bac or without	Bac+2	Bac+3	
Sample	44.5	21.0	34.4	
National population	62.9	14.4	22.4	

Source: For the national population, the most current available data of INSEE was referred respectively.

Note: The annual household income was initially answered with a 10,000-euro scale but later adjusted according to household structure by INSEE. For the sake of statistical convenience, the respondents were then grouped into three income categories of Observatoire des Inégalités (2019), which consists of high-income (upper 20%), middle-income (in-between 30–80%) and low-income groups (bottom 30%). For example, in a household of 'parents with two children', the segmentation is as follows: high-income (70,000 euros >), middle-income (70,000–40,000) and low-income (>40,000).

7.1.3 Social Representation Theories

The Web questionnaire comprised three different sections: two questions about the citizens' naïve representation of gastronomy, 24 questions about its concrete eating model or practice dimensions (e.g., where, when), and several questions about their socio-demographic status.

To elucidate the first part of gastronomy, I employed a theory of social representations originally developed by S. Moscovici in the 1960s, which directly or indirectly inspired the sociology of eaters (Lahlou 1995; Corbeau & Poulain 2002; Fischler & Masson 2008). Social representations are similar to images, ideas and concepts of given objects in our everyday language, but its epistemological status needs to be specified (Moscovici 2001). It refers to (1) a common sense of ordinary people, which is (2) dynamic, (3) specific and (4) allows for interpersonal differences, as opposed to the static and general nature of Durkheim's collective representation. It is named 'social' because it (5) also refers to shared knowledge within given social groups and (6) functions as a social environment (not a uniquely mental image often seen in orthodox psychology) that (7) frames our behaviours.

Three major approaches to social representations, namely the process of genesis, the structure and the operating function in communication, have been proposed (Rateau et al. 2012), among which the 'structure' of social representation (i.e. gastronomy) is the focus of this analysis. Abric (1993) specified that such structure is composed of two sub-systems. The 'central core (system)' is stable and not very sensitive to the immediate context or individual differences, generating the major signification of the given representation. In contrast, the 'peripheral system' is more flexible and bears contradictions, allowing for adaptation to concrete reality and content differentiation, and, therefore, protects the central system.

Among several structural approaches to social representations (Monaco et al. 2017), the quantitative method developed by Lahlou (1995) was insightful, because, together with his colleagues (Mathé et al. 2014), he analysed the content and evolution of the social representation of *bien-manger* (eating well) for over 30 years with a lexical analysis programme (known as ALCESTE). In my analysis, this method was applied with a free programme of essentially the same function (called IraMuteq) to elucidate the social representation of gastronomy. The detailed protocol was described in Appendix. The identified semantic classes from respondents' open-ended answers to the question 'what is gastronomy for you?' are regarded as the nuclei of given representation (that include the core and peripheral systems). In other words, to elucidate the French public definition of gastronomy can be equated to identifying these semantic classes.

7.1.4 *Evaluative Dimensions of Eating Model*

While the semantic nuclei of gastronomy can be elucidated in the first section, an additional approach should be taken to identify 'pre-discourse' (Lahlou 1995, p.298), in other words, the content perceived as given and not easily verbalised by the population in question. Therefore, this second section aimed at exploring such pre-discursive elements. To achieve this, the selection of questions was informed by major perceptual gaps among the French identified in the exploratory interview as well as previous literature on the modèle alimentaire (eating model) (Herpin 1988; Poulain 2002a, 2002b) and the practice of eating (Warde 2016). Although these two scholarships have differing intellectual sources – the former from the sociology of eaters (Morin, Fischler, Corbeau, Poulain) and the latter from classical theories of practice (Giddens, Bourdieu) – they share the mission to capture the totality of eating practice by going beyond the traditional view in which eating was considered merely as a combination of consumed foodstuffs or cultural expression.

Poulain (2002b, p.57) identified six dimensions for evaluating the eating model: spatial, temporal, intake structure, social environments, logic of choice and corporal position. Similarly based on a wide-ranging source of studies in anthropology and sociology, Warde (2016, pp.52–79) identified three elementary dimensions that can define the eating practice: events/occasions, food/menu/dish and incorporation/body techniques. Informed by these two types of comprehensive catalogue, 24 questions were selected to explore the total dimension of the gastronomic eating model (for the detail of selection of question items, see Ueda & Poulain 2021, as well as Appendix).

7.2 Gastronomy as Social Representation

The analysable corpus was composed of 430 texts (an average of 9.21 words per text), a total of 4,045 words of 584 different lexical forms, and five semantic classes with 72.9% classification rate (which was better than the acceptable 70%).

7.2.1 Art of Cooking Quality Food Ingredients

The first semantic class consisted of frequently used words such as *cuisine* and *quality*. Some of the characteristic texts are described below in the order of relevance. Bolded words represent the characteristic ones to a given class ($p < 0.05$). The number on the left simply refers to the assigned number to each text (i.e., respondent):

318 *le plaisir de bien manger des* ***aliments*** *de bonne* ***qualité*** *et différents de la cuisine de tous les jours* (the pleasure of eating well on good quality food and differently from the everyday cuisine)
405 *de bons plats préparés avec des techniques des* ***cuisiniers professionnels****, des* ***aliments de qualités supérieures*** *qui font le prix élevé du plat* (good dishes prepared with the techniques of professional chefs, food ingredients of superior quality that make the highest price for the dishes)
138 *l'****art*** *de* ***cuisiner*** *et de* ***sublimer*** *les* ***produits alimentaires*** (the art of cooking and enhancing the food products)

Thus, this semantic class can be named the 'art of cooking quality food ingredients'. This culinary art referred not only to the mere use of quality food ingredients but also to the highly professional skills to enhance such ingredients. The fact that more respondents had high education levels ($p < 0.05$) in this class implied that this aspect of gastronomy is viewed as a 'difficult art' that necessitates a high level of experience, knowledge and education.

7.2.2 Culinary Traditions Representing the Communities

The second class was composed of frequently used words such as *culinary*, *tradition*, *taste*, *sophisticated* and *country*. Some characteristic responses include:

161 *des plats* ***traditionnels représentatifs*** *de la* ***culture*** *et de savoir-faire* ***culinaire*** *d'un* ***pays*** (traditional dishes representing the culture and culinary know-how of a country)
341 *la maîtrise d'une* ***culture culinaire*** *précisée des* ***plats*** *de qualités et très* ***goûteuse*** (the mastering of culinary culture characterised by quality and very tasty dishes)

323 *l'**ensemble** des habitudes alimentaires et **traditions culinaires** d'un group **donné*** (food habits and culinary traditions of a given group)

This semantic class, therefore, can be described as 'culinary traditions representing the communities'. The communities included diverse levels such as countries and regions, each of which has sophisticated its unique traditions of culinary skills, recipes and eating habits. Eventually, this class was very close to the definition of intangible cultural heritage (UNESCO 2010; Csergo 2016).

7.2.3 Grand Chefs in Michelin-Starred Restaurants

The third semantic class was composed of frequently used words such as *chef*, *grand* and *restaurants*. The characteristic texts are described below:

229 *le **plat** que l'on mange en **grand restaurant cuisiné** par un **grand chef** souvent trop cher et bon* (an excellent dish that we eat in a famous restaurant, cooked by a famous chef and is often too expensive but tasty)
271 *la **cuisine** de **terroir** de **grands chefs étoilés*** (the cooking of terroir by famous Michelin-starred chefs)
343 *un art de la **cuisine servie** dans les **restaurants étoilés*** (the art of cooking served in the Michelin-starred restaurants)

This semantic class can, thus, be referred to as 'grand chefs in [Michelin] starred restaurants'. This class resonated with the first semantic class because these Grand chefs can practice this 'art of cooking'. Moreover, the fact that this class was characteristic to the group of low education level ($p < 0.05$) added more evidence to the correlation between education level and the elitist view of gastronomy as already observed in the first class.

7.2.4 Eating with Pleasure

The fourth semantic class was composed of frequently used words such as *eating* and *pleasure*. The characteristic texts are described here:

416 *le **plaisir** de bien **manger*** (the pleasure of eating well)
270 *des produits de qualité que j'ai **plaisir** à **manger** dans des restaurants* (quality products that I have the pleasure of eating in the restaurants)
381 ***aimer manger** de bon plat cuisiné* (to enjoy eating good cooked dishes)

This semantic class, thus, can be described as 'eating with pleasure'. In contrast to the third semantic class where chefs were the subject of practice, the agency aspect of the respondents as eaters was emphasised in answers such as I enjoy eating and [I] like to eat. Given that no characteristic group was observed in this class, in contrast to the other three classes, eating with pleasure can be regarded as being shared by the French public, regardless of their socio-demographic status.

7.2.5 Expensive Meals

The fifth semantic class was composed of frequently used words such as *meal* and *expensive*:

189 *un* ***repas*** *haut de gamme et* **cher** (a high-end and expensive meal)
252 *un* ***repas*** *fait pour la* **riche** (a meal made for the rich)
396 *l'art de faire des* **petites assiettes** *avec des produits super* **chers** (the art of cooking small dishes with very expensive products)

This semantic class, therefore, can be described as 'expensive meals'. While classed as gourmand and sophisticated, this gastronomic meal was considered (too) expensive and exclusive to the rich, which strongly indicated the elitist dimension of gastronomy. In supporting this vision, those from a low-income level were characteristic of this class ($p < 0.05$).

7.2.6 Coexisting Plural Perspectives

To summarise the first result, gastronomy in French citizens' representation was composed of five semantic nuclei: the art of cooking quality food ingredients, culinary tradition representing the communities, grand chefs in Michelin-starred restaurants, eating with pleasure and expensive meals. As gastronomy is often paraphrased as 'the art of eating well', the result is very consistent the representation of eating well, based on eight semantic nuclei (subject: eating what we like and eating enough; object: starter, warm main dish, cheese and dessert and small dishes; modality: convivial and restaurant; finality: balance and not too much fat and sugar) (Lahlou 1995).

Nevertheless, it is worthwhile to note the nuanced differences between gastronomy and eating well. First, gastronomy is more characterised by the absence of physiological or nutritional discourse (such as eating enough [fill the stomach] or not too much fat and sugar in eating well) and the resultant emphasis of its cultural dimension (e.g., the art of cooking and culinary traditions in gastronomy). Second, gastronomy certainly contains some elitism, as the influence of education or income level was observed in three of the semantic classes. To quote some relevant responses from the exploratory interview:

> What can I say about gastronomy? I'm not the right person to answer this, you should ask grand chefs.
>
> (Madame, 40s, owner of a small cafe in Tours)

> Excellent taste, pleasant atmosphere, beautiful plates. You can find these only in gastronomic restaurants. And not only this, some clients have purchasing power worthy of these restaurants.
>
> (Madame, 50s, housewife in Toulouse)

However, it would not be appropriate to reduce gastronomy merely to such an elitist dimension, as it was also confirmed that, aside from such limited perspectives of Grand chefs and restaurants, the agency aspect as eaters who enjoy eating as well as community members who carry traditional food cultures was also expressed by the French public of any socio-demographic status. In other words, these coexisting plural perspectives of chef-artists, eaters and community members have made it more complex to articulate the definition of gastronomy. This is the third difference from eating well for which the subject is always eaters themselves. Again, while the elitist dimension constitutes an aspect of gastronomy, we should not dismiss other dimensions such as its agency or its identity-building aspect for certain professionals, for social representations function as the 'reference point' that allows these actors to take social positioning (Doise et al. 1993) (Figure 7.1).

> Gastronomy…it's the pleasure of eating! It's a chance for France. This idea is unique, but France has yet to succeed in exporting this idea to the world. While the Americans eat for the appetite, the French eat for pleasure.
>
> (Monsieur, 50s, wholesaler in Toulouse)

> Gastronomy is all about my life. I have been working in various places, from a famous restaurant to a traditional restaurant. But all of these experiences are gastronomy for me.
>
> (Monsieur, 50s, wine retailer in Lyon)

7.3 Gastronomic Eating Model

Hereafter, I will discuss the results of the second section in the web-based survey asking the composing elements of gastronomic eating practice. The question

Figure 7.1 Interview with Professionals in Toulouse Wholesale Markets

Source: Photographed by the author. Scenes of the direct negotiation between wholesalers and retailers and the loading of vegetables and fruits for the market.

items can be referred to in the corresponding tables (Table 7.2). I highly recommend that those who wish to delve into the depths of the French gastronomic world read each of the following sections; however, since a great deal of quantitative data is presented here, it is also possible to jump to the 'Commonality and Social Difference' section at the end and, whenever necessary, come back to the corresponding sections for empirical justification.

7.3.1 *Spatial: Gastronomy-In and Gastronomy-Out*

For its spatial dimension (Q3), 87% of the total respondents considered gastronomy as a practice that can be performed inside and outside the home, so, hereafter, I shall operationally refer to them as gastronomy-in and gastronomy-out.

> We have several types of gastronomies. There is the gastronomy of Grand chefs, traditional and French gastronomy, and familial gastronomy.
>
> (Madame, 60s, retailer in Toulouse)

However, comparatively speaking, gastronomy-out was considered more appropriate than gastronomy-in, since only a few answered that a gastronomic practice takes place more at home, but also at restaurants or uniquely at home. This restaurant-dominant vision was more characteristic of the young and the women. The more reinforced version was observed among the 11.9%, particularly those with low incomes, who relegated gastronomy uniquely to the restaurant realm. The reason (Q4) for this exclusion of gastronomy-in was explained as being due to gastronomy being a practice reserved for chefs and restaurants (n = 21), requiring skills, know-how (n = 10) and special equipment (n = 5) that cannot usually be found at home, and as too complex (n = 6) and expensive (n = 4). These reasons resonated with some semantic dimensions identified in the first analysis (art of cooking, grand chefs and expensive meals).

7.3.2 *Timing: Most Typically as a Festive Dinner but also as Non-Festive Meals*

For its temporal dimension (Q3), around 80–90% of respondents considered a holiday dinner as most typical for practicing gastronomy. Generally, gastronomy was regarded more appropriate in the following order: as for the week, holidays, weekends and weekdays; and as for the day, dinner, lunch and breakfast. In terms of the socio-demographic difference, the younger perceived a wider range of timing as appropriate for its practice than the older did.

This result is consistent with UNESCO's definition (2010) of festive meals, but it should also be noted that more than half of the respondents considered not-necessarily festive meals, even during weekdays and weekends, as gastronomic. This indicated that gastronomy is not reduced only to festive meals but rather can be practiced for a wider range of occasions if it satisfies certain conditions, which we explore in the following sections.

Table 7.2 Space, Timing and Persons to Share with and Duration of Gastronomic Meals

Q1. First, if I say 'gastronomy' to you, what are the five words that spring to mind?

Q2. This time, if I ask you to define 'gastronomy'... Gastronomy, in your own words, is...?

Q3. Generally, for you, gastronomy can be practiced...?

Uniquely at restaurant	More at a restaurant, but also at home	As much at a restaurant as at home	More at home, but also at a restaurant	Uniquely at home
11.9	40.2	46.8	1.1	0
LI 16.7 > HI 6.3, MI 7.4	Y 47.0 > O 31.3	M 53.3 > F 40.7 O 56.7 > Y 37.1		

Q4. To return to your previous question, why is home [restaurant] not concerned in gastronomy?

Q5 [Q6]. For you, what times are important in gastronomic meals at home [restaurant]?

	Breakfast	Lunch	Dinner	Appéritif, snack
Holidays	7.5 (7.3)	66.6 O 81.0 > Y 56.8, M 62.3 (63.5) (F 68.1 > M 58.5)	87.6 F 91.8 > M 83.3 (79.9) (M 84.3 > O 70.9)	7.8 F 12.9 > M 2.6 (3.7)
Weekends	16 Y 24.3 > M 11.0 (8.4)	58.3 (62.6)	68.1 (78.1)	8.0 F 10.0>M 4.7 (3.7)
Weekdays	9.3 (7.8) (Y 14.4 > M 4.7)	24.6 (50.5) (Y 57.6 > O 41.0)	57.3 (54.8) (Y 62.1, M 59.9 > O 41.0)	8.5 (7.1) (MI 10.4, LI 7.5 > HI 0.0)

Q7 [Q8]. With whom do you share gastronomic meals at home [restaurant]?

My close family	My extended family	My friends	I can practice it alone
87.3 (84.7)	36.8 MI 45.6 > LI 28.4 (31.5) (HI 44.0 >LI 25.4)	64.8 F 72.2 > M 57.3 (61.6) (HI 78.7 > MI 62.2, LI 55.7)	6.2 (3.7) (Y 6.8 > M 1.2)

Q9. During the gastronomic meal, a minimum of how long is needed for eating (except for the time of preparation)?

Median: 90, quartile range: [60, 120] O 120 > Y 90, M 90

(Median: 120, quartile range: [70, 120] F 120 > M 90) Out > In

Note: F: female, M: male, Y: young, M: middle, O: old, HI: high income, MI: middle income, LI: low income, HE: high education, ME: middle education, LE: low education. The results regarding gastronomy-out are described in the parenthesis.

7.3.3 Persons to Eat with: Indispensable Presence of Intimate Others

More than 60% regarded gastronomy as the practice of eating together with close family or guests (such as close friends) (Q7–8). Although eating alone was also considered gastronomic by a small portion of the respondents, they answered that it is because they like to cook and try new recipes alone for future guests or family members. This clearly indicated that the presence of intimate others is indispensable for practicing gastronomy. Regarding the socio-demographic difference, those with higher income were more likely to demand special or unusual moments to spend with those beyond their close family (such as relatives and friends).

> In one word, gastronomy is like this. You should take note because it's very important – *Mettre les petite plats dans les grands!*
>
> (Monsieur, 50s, retailer at marché in Toulouse)

7.3.4 Meal Duration: Relaxing Time of 90–120 Minutes

The meal duration (Q9: median) was 90 mins for gastronomy-in and 120 mins for gastronomy-out. The older and the men preferred a longer duration for gastronomy-in. Since an average daily meal duration for a French family is 95 mins in total (according to the data of INSEE in 2010), gastronomy necessitates meal duration equal to this average or more; thus, again, it should be considered a relatively special or unusual practice (Figure 7.2).

Figure 7.2 Interview with Livestock Producers in a Suburban Area of Toulouse

Source: Photographed by the author. Taking a relaxed time at weekday lunch served with Italian polenta, oxtail soup and wine.

I can't believe you guys have only one hour for lunch in the conference on gastronomy? Give me phone numbers, and I need to say something to your profs!

(Monsieur, 40s, restaurant owner during the conference on food cultures in Tours)

7.3.5 *Place for Procuring Foodstuffs: From Producers and Marchés*

About 70–80% of the respondents regarded farmers or marchés (outdoor markets) as the best source for procuring gastronomic foodstuffs (Q10). In practicing gastronomy-in, they evaluated these means as favourable for procurement, which include building relationships with farmers and retailers, while more commercialised means, such as supermarkets or discount shops, were evaluated as unfavourable.

Even if it's difficult during weekdays, on weekends I go to a marché to buy fresh chicken from producers…I can't do gastronomy with Carrefour. It's impossible.

(Madame, 50s, independent insurance consultant in Toulouse)

7.3.6 *Places for Eating Out: Michelin-Starred and Traditional Restaurants*

Over 70% of respondents considered Michelin-starred or traditional restaurants as appropriate for gastronomy-out. Other places (such as bistros and brasseries) were regarded as appropriate only by 20–30%, among which there were more respondents with a high education level.

7.3.7 *Quality of Food/Restaurants: Inspiration and Locality/Tradition*

The 'quality' of food and restaurants was analysed using the convention theory, in which the quality can be understood as being coordinated within different worlds (*cités*) of order (see Appendix). Overall, the *cité* of inspiration and the domestic cité were regarded as the most important settings for defining the quality of both gastronomy-in (Q11: foodstuffs) and gastronomy-out (Q13: restaurants). In contrast, the commercial and industrial *cités* were evaluated as relatively inferior, and the higher income group expressed this tendency more evidently (Figure 7.3 and Table 7.3).

Cité of Inspiration – When buying our vegetables, he [the chef] already knows what to cook. The flavors are already spreading in his mouth; it is like an 'artist' in a true sense… I think a person like that can be called the genuine gastronome.

(Monsieur, 50s, retailer at marché in Toulouse)

Figure 7.3 Interviews with Artisans and Cassoulet (Regional Recipe) in Toulouse
Source: Photographed by the author. 'Quality' was repeatedly emphasised by these artisans (bakery, butcher, etc.).

> Domestic Cité – In fact, even high gastronomy for the elite shares the same origin with the popular tradition transmitted from our ancestors of each region. Even Paul Bocuse learns how to cook from his grandmother.
> (Madame, 40s, independent insurance consultant in Toulouse)

7.3.8 *Recipes: Regional Hope Recipes and Fish-Based Restaurant Recipes*

For gastronomy-in, *boeuf bourguignon* (n = 34), *choucroute* (n = 24) and *couscous* (n = 22) were considered as most appropriate recipes. It is interesting to note that they chose regional recipes originally based on local and popular cultures (of Bourgogne and Alsace), rather than aristocratic dishes, with the corresponding reasons (such as traditional and regional). Moreover, this result was consistent with the 'culinary tradition' dimension identified in the first analysis.

For gastronomy-out, only the names of foodstuffs, not of recipes, were mentioned in the answers. This might be partly due to the increasing complexity of the appellation of restaurant dishes under the Nouvelle Cuisine movement. These gastronomic foodstuffs included fish in general (n = 28), lobsters (n = 18) and scallops (n = 12), with the reasons such as [they are] difficult to prepare at home, luxurious and expensive and special and unusual. It is also interesting that many respondents included couscous as a gastronomic recipe, which was imported from Maghreb immigrants in the late 20th century. The integration of such regional and exotic cuisines constitutes solid evidence of 'gastronomic decolonisation' (Poulain 2011) (Table 7.4).

7.3.9 *Meal Structure: Appéritif, Starter, Main, Cheese, Dessert and Wine*

More than half of the respondents listed appéritif, starter, main dish, cheese and wine as essential meal components both for gastronomy-in and gastronomy-out

Table 7.3 Gastronomic Quality of Foodstuffs and Restaurants

Q10. For you, to prepare gastronomic meals at home, the products should be purchased from…?

Self-produced	Farmer	Marché	Small retail shop	Super/hypermarket	Discount shop
47.7	72.0 F 77.8 > M 66.1	77.7	43.0 HI 57.7 > LI 35.8	33.4	8.3

Q11. What criteria are important for the principal products when preparing gastronomic meals at home?

Regional, traditional	Original, passion (of producers)	Social advantage (environment, animal well-being…)	Attractive price	Effective (to cook)	Famous (label)
3 (2.93)	3 (2.98)	3 (2.74) F 3[2, 4] > M 3[2, 3] Y 3[2, 4] > M 3[2, 3], O 3[2, 3] LI 3 > HI 2	3 (2.53) LI 3 > MI 2, HI 2	3 (2.64) LI 3 > MI 2	3 (2.77)

Importance : Inspiration = Domestic > Opinion = Civil = Commercial = Industrial; Opinion = Civil > Commercial

Q13. What criteria are important when you choose the restaurant for a gastronomic meal?

Regional, traditional	Original, passion (of chefs)	Social advantage (environment…)	Attractive price	Effective (to cook)	Famous (label)
3 (2.94)	3 (3.26) F (3.35) > M (3.16)	3 (2.73) F 3[2, 4] > M 3[2, 3] Y 3 > O 2.5	3 (2.71) F 3 (2.78) > M 3 (2.63) Y[2, 4] > O[2, 3], M[2, 3] LI(2.90) > MI(2.50), HI 2	3 (2.56) Y 3, M 3 > O 2 LI 3>HI 2	3 (2.81)

Importance: Insp. > Dom. = Opi. = Fam. = Com. = Ind.; Dom. > Opi. = Civil = Com. = Ind.; Opi. = Civil = Com. > Ind.

Q12. At which restaurant can you practice gastronomy?

Michelin-starred restaurant	Traditional restaurant	Bistro	Brasserie	Café, bar	Cantine (school…)
80.1	73.1	21.7 HE 29.1 > LE 15.3	25.8: HE 32.4 > LE 20.0	4.3 M 7.6 > O 1.5	3.9

Note: For Q11 and Q13, 1–4 indicate 'very important (4)', 'important (3)', 'slightly important (2)' and 'not important (1)'. Although representative values are median and in the quartile range, to visualise the statistical difference, the average value in the parenthesis is also described.

Table 7.4 Recipes, Meal Components and Pleasure of Gastronomy

Q14. In your opinion, which dishes represent gastronomic meals? For what reasons do you associate with gastronomy?

Gastronomy-in :		
Boeuf bourguignon	34	Traditional (n = 10), tasty and rich in flavour (n = 10), copious (n = 5)
Choucroute	24	Copious (n = 8), traditional (n = 7), regional (n = 5)
Couscous	22	Tasty (n = 6), copious (n = 5), healthy and rich in vegetables (n = 5)
Blanquette de veau	15	Traditional (n = 6), tasty and rich in flavour (n = 4), easy to cook (n = 3), sophisticated (n = 3)
Cassoulet	12	Tasty (n = 5), traditional (n = 2), copious (n = 2)
Pot au feu	12	Traditional (n = 4), tasty and rich in flavour (n = 3)
Lasagne	12	Tasty (n = 5), copious (n = 3), easy to cook (n = 3)
Gastronomy-out:		
Poisson	28	Difficult to cook and professional (n = 7), tasty (n = 6), sophisticated (n = 6)
Homard	18	Luxurious and expensive (n = 8), special (n = 7), tasty (n = 2), difficult to cook (n = 2)
Fois gras	16	Special (n = 5), sophisticated (n = 4), tasty (n = 3)
Cassoulet	12	Tasty and rich in flavour (n = 5), traditional (n = 2)
Noix de saint jacques	12	Special (n = 3), tasty and rich in flavour (n = 2), sophisticated (n = 2)
Viande	11	Tasty and rich in flavour (n = 2)
Couscous	10	Tasty and rich in flavour (n = 5)
Magret de canard	10	Traditional (n = 2), sophisticated (n = 2), tasty and rich in flavour (n = 2)

Q15 [Q16]. In your opinion, which indispensable elements constitute gastronomic meals at home [restaurant]?

Appéritif	Soup	Hors d'oeuvre	Starter	Main dish
56.7 M 62.5 > F 51.0	7.5	32.4	74.1 MI 83.2 > LI 70.5	90.9 O 95.9 > Y 85.6
(61.4) (LE 68.7>HE 49.0)	(9.1) (Y 15.2 > O 6.0)	(37.2)	(78.5)	(89.7) (O 94.8 > Y 85.6)

Second main dish	Cheese	Dessert	Wine	Digestive
16.8	64.8 O 76.9> Y 48.6, M 66.9	83.7 O 89.3 > Y 76.6	72.5 O 90.9 > Y 55.9, M 70.1 MI 83.2 > LI 64.2	25.9
(27.6) (LE 22.9 > ME 7.2)	(56.8) (O 68.6 > Y 47.7, M 54.7)	(84.2)	(71.0) (O 84.3>Y 62.1, M 67.4)	(26.9)

Q17 [Q18]. In your definition of gastronomy at home [restaurant], the notion of pleasure is important because…?

It satisfies my appetite	I enjoy the taste of dishes	It is a moment of exchange with others who share the table	I feel a sense of privilege
24.4 Y 36.0, M 26.0 > O 11.6	67.9 O 76.0 > M 61.7	70.7 O 80.2 > M 65.8	12.7
(24.7) (Y 33.3 >O 17.2)	(74.0)	(58.0) (O 80.6 > Y 66.7; HI 69.3 > LI 53.1)	(37.0)

(Q15–16). This is more demanding than the structure of 'eating well' (starter, main dish, cheese and dessert) (Lahlou 1995) and more consistent with 'a course composed at least of four dishes' in UNESCO's definition (2010). In the case of the socio-demographic difference, the young and the low-income groups tended to simplify the meal structure. It was implied that these six components serve as a reference for gastronomic meals and other elements can be added/omitted depending on the socio-economic situation of each section of the population (Figure 7.4).

7.3.10 *Pleasure: Gustative and Socialised Pleasure*

More than half of the respondents appreciated the pleasure of eating for gustative (taste of dishes) and socialisation (conviviality at table) reasons (Q17–18). With regard to the socio-demographic difference, the younger prioritised physiological pleasure (satisfying appetite), while the older placed more emphasis on socialised pleasure. Particularly in terms of gastronomy-out, 37% enjoyed a sense of privilege, that is, the social distinction dimension at restaurants that has played an important role in the history of gastronomy (Aron 1973).

> Gustative pleasure – There is no gastronomy in normal cafes and brasseries. They are the industries for satisfying appetite, but we are different. We do our jobs for the pleasure of eating good food.
>
> (Monsieur, 30s, artisanal butcher in Toulouse)

> Socialised pleasure – Gastronomy is the art of living. It is [the art of] having pleasure, having a good time, and sharing meals with friends and family.
>
> (Madame, 40s, secretary at research institute in Paris)

Figure 7.4 Interview with an Ordinary Family in Périgueux

Source: Photographed by the author. The family comprises a couple, who are a clerk in a private company (husband) and a junior high school teacher (wife), with a daughter who is a nutritionist working in a hospital.

Note: The composition of the dinner was as follows: *lillet blanc*, regional apéritif in Bordeaux; foie gras with orange confiture as starter; confit de canard with sautéed potatoes as main dish; goat's cheese with green salad; walnut pancake with fresh orange as dessert; and a variety of liquors as digestif. These foodstuffs were procured from the marché and professional retailers.

7.3.11 *Priority: Building Harmonious Relationships with Others*

Among the abovementioned dimensions, 'persons to eat with', 'timing', 'pleasure' and 'recipes' were prioritised both for gastronomy-in and gastronomy-out (Q19–20). If combined with the previously identified content of each dimension, gastronomy can be characterised as the practice of eating together with close family and friends at appropriate times, such as a holiday, with the pleasure of socialisation as well as the taste of regional and traditional recipes. Again, building harmonious relationships with others was emphasised.

7.3.12 *Values of Gastronomy: Pleasure and Tradition*

Thus far, each dimension of the gastronomic eating model was concretely characterised, but we also asked how such gastronomy was valued by the French public (Q21). More than 80% valued gastronomy positively because it brings conviviality (n = 93), the pleasure of tasting good dishes (n = 73) and pleasure in general (n = 63). This result is consistent both with the 'eating with pleasure' class (Q2) and with the gustative and socialised pleasure (Q17–18). The other dominant reasons were that gastronomy is traditional (n = 52) and part of the French or regional heritage (n = 32), which is consistent with the 'culinary tradition' class (Q2). In contrast, gastronomy was valued unfavourably by more than 10% because, for them, simple meals are enough (n = 13), gastronomy is too expensive and luxurious (n = 11), and they are not interested or do not like such cuisine and form (n = 6).

7.3.13 *Obstacles in Practicing: Not the Lack of Interest or Motivation*

The obstacles to practicing gastronomy were explored (Q22, 24). While technical and financial constraints were considered as equally significant obstacles for gastronomy-in, the lack of interest or motivation was not. In the case of gastronomy-out, finances, time and technical or motivational constraints were regarded as the most significant obstacles in descending order. The fact that financial constraint was the largest obstacle at the restaurant is consistent with the answers such as too expensive among those who negatively valued gastronomy (Q21). However, it should be noted that non-motivational (technical, financial, time) constraints hinder the French from practicing gastronomy, which, in other words, implies that interest and motivation for gastronomy are widely shared among the French population.

> Technical constraint – If I were asked to do gastronomy at home, I would never be able to do that. Because I would need to use quality food, prepare tasty dishes, follow certain rules…I don't know all of them.
>
> (Madame, 20s, staff member at food-related non-profit association in Paris)

> Time constraint – Spending time on eating is a simple act but has become a sort of luxurious act today. I think the reason why we can't practice gastronomy anymore is not the financial matter, but the matter of our lifestyle.
>
> (Monsieur, 30s, owner of a local canteen in Tours)

> Financial constraint – Are we really saying gastronomy is festive family meals [in the UNESCO definition]? For me, I've never thought that. Even so, it is for normal people. Because, the popular class, like me, or rather, the working class like me…we can't go to gastronomic restaurants that frequently. It's too expensive.
>
> (Madame, 20s, unemployed in Toulouse)

7.3.14 *Practice Frequency: Actually 2–3 Times/Year, Ideally 6 Times/Year*

The respondents actually practiced gastronomy 2–3 times/year and ideally wished to do so 6 times/year (Q26: median). The breakdown of the result also showed that a large portion of the respondents could practice gastronomy several times and more per year. This level of frequency is consistent with 'festive meals' (UNESCO 2010) and the previous result (Q5) in which gastronomy can be practiced even for lunch and dinner during weekdays and on weekends. Regarding the socio-demographic status, those with a higher income and education level practiced it more frequently, and this tendency is consistent with the result (Q24) in which the financial constraint was considered as the first or second largest obstacle (Table 7.5).

7.4 Commonality and Social Difference

To summarise these results (Q3–26), the core content of the gastronomic eating model that was commonly shared by the French population can be described as follows:

> The core of the gastronomic eating model – Gastronomy is a 'practice that is possible to perform both inside and outside home', so it is not the privilege exclusive to chefs and restaurants. It is the 'practice of eating together with intimate others (family and friends)', so it cannot be performed alone. It is also the practice that 'takes place most typically as a holiday dinner but not necessarily limited to such festive timing' and that necessitates 'regional, traditional or original foodstuffs' to prepare 'special and traditional recipes', with 'the pleasure of tasting such dishes and socialisation with the others'. It is preferable to 'procure these foodstuffs from farmers or marches', and if eating out, it is preferable to 'enjoy these recipes at Michelin-starred or traditional restaurants'. To obtain such pleasure, the meal should be composed of 'appéritif, starter, main dish, cheese, dessert and wine' as a minimum, and a 'relaxed time of more than 90–120 minutes should be spent at the table'.

Table 7.5 Priority, Values, Obstacles and Frequency of Gastronomic Practice

Q19 [Q20]. When having gastronomic meals at home [restaurant], what importance do you attach to the following elements?

Timing	*Persons to eat with*	*Meal duration*	*Place of procurement/ restaurant*
3 [1,5] (3 [2,5])	2 [1,4] (2 [1,3])	7 [5,8] F 7 > M 6 (6 [4,7]) (F 6>M 5; O 6>Y 5)	5 [4,7] (3[2,4])
Quality of food 3 [2, 4] Y 4 > O 3 (-)	Meal structure 6 [4,7] (5 [4,6]) (Y [4,6], M [4,6] > O [3,6]) (LI [4, 6] > MI 5 [3, 6])	Recipes 5 [3,7] Y 6 >M 5, O 5 (4 [3,6]) Y5 > O 4)	Pleasure 4 [3,6] F 5 > M 4 (4 [2,6])

Q21. Considering the gastronomy you have described until here, is it important for you? Can you explain why it is important or not important?

Important	85.8	Conviviality and the pleasure of exchange with others (n = 93), the pleasure of tasty dishes and encountering new flavours (n = 73), like to eat and the pleasure of eating [in general] (n = 63), [French or regional] tradition and heritage (n = 52), special and unusual (n = 32)
Not important	12.3	Simple meals are enough (n = 13), too expensive and luxurious (n = 11), no interest, don't like cooking or the form (n = 6), other things (such as health) are more important (n = 5), I don't or haven't practiced it (n = 4)
No answer	1.8	-

Q22 [Q24]. For you, what are the obstacles to practicing gastronomic meals at home [restaurant]?

Time constraint	Financial constraint	Lack of savoir-faire or skill	Lack of interest or motivation
3 [1, 4] (3 [0,3]) (Y 3[2, 3] > M 3[0, 3] > O 2)	3 [1, 4] (4 [3, 4])	3 [2, 4] (1 [0, 2])	1 [1,2] M 1(1.60) > Y 1 (1.30) (1 [0, 2]) (LI 2 > MI 1)

Q26. Last year (in 2018), approximately how many times did you manage to practice gastronomic meals at home or in a restaurant?

Actual: 3 [1, 8] (2 [1, 5]: HI 5 > MI 2, LI 2; HE 3 > ME 2, LE 2)

Ideal: 6 [4, 20]: LI 6 [4, 20]> MI 5 [3, 12] (6 [3, 12]: HI 10 > MI 6, LI 5; HE 8 > LE 5)

Note: For Q19–20, the ranking evaluation of 8-grade for gastronomy-in and 7-grade for gastronomy-out, in which smaller values represent more importance. For Q22 and Q24 the ranking evaluation of 4-grade, in which larger values represent larger obstacles. The personal reasons for these answers were also asked (Q23, 25) but not included here.

Although this core part of gastronomy was widely shared among the French population of any socio-demographic status, after further analysis, a socio-demographic-dependent tendency was also observed in conceptualising such gastronomy. Regarding the gender gap, the women tended to perceive gastronomy as more 'unusual or special' by prioritising the practice at restaurants (Q3), festive meals (Q5–6), the presence of guests (Q7), longer meal duration (Q3) and originality of restaurants (Q13). This might be the converse effect of the fact that women are still carrying heavier roles in providing 'usual or everyday' meals than men are. With regard to the generational gap, older people were more demanding in practicing gastronomy at home (such as Q5: festive lunch, Q9: longer meal duration, Q15–16: cheese and wine, Q17–18: the pleasure of taste and socialisation), while for the young, these norms tended to be weakened – or 'destructured' (Poulain 2002a) with a multitude of new dietary norms.

Regarding the social class (income and education) gap, the lower social class tended to be less demanding for certain elements (Q7–8: eating with relatives and friends, Q18: wine), have a narrower vision of gastronomy (Q10: exclusion of small retailers, Q12: exclusion of brasseries), and practiced it less frequently (Q26). Nevertheless, the difference that depended on social class was observed only in such limited ways, so it can also be said that the effect of social class was not as significant as indicated in previous literature (Bourdieu 1979).

7.5 Concluding Remarks

For the French public, gastronomy has a long history and occupies a special relationship with them in the national narrative. Two literary veins on gastronomy were identified: gastronomic criticism and academic analysis. Even if the scholarly discourse has influenced critical literature since the 1960s, these two veins are very different. The former is normative, whereas the latter is analytical and interpretative. This work is explicitly situated in the second perspective. Using a method combining a series of exploratory interviews and an online questionnaire, this analysis completes and extends the study of audiences by refining the understanding of the implicit definition of gastronomy for the French population.

Although differences are discernible, there is an important 'central core' of content that is relatively independent of the people's social position. For the overwhelming majority of French people, gastronomy is associated with the pleasure of eating, and this pleasure is not reducible to the products, the recipes or the taste experience itself. It is a shared and socialised pleasure. This finding is in line with a long series of academic works that has enlightened the connection between pleasure and conviviality in French food culture (Fischler 1990; Corbeau 2008; Dupuy 2013). At the same time, the finding challenges Bourdieu's thesis on the autonomy of working-class taste and, instead, supports the thesis of transcendental transmission of gastronomy (Elias and Poulain).

On the other hand, it is also important to emphasise that, behind this consensus, the forms of gastronomic practices could vary depending on social positions. In addition to the effects of social positions on the peripheral content of gastronomy, one interesting finding is that discursivity (i.e. the ability to verbalise the gastronomic experience), characteristic of classical and normative gastronomic criticism, was not observed in the lay population from this research set-up. Nevertheless, it is too soon to attribute this lack of discursivity to the fixed characteristics of eating practice by the socially disadvantaged (as advocated by Bourdieu); rather, it would be more pertinent to interpret such social variation as the gap between the gastronomic 'ideal' and its 'actual' practices of contemporary eaters, which eventually justifies educational intervention to develop people's expressive capability.

Further research should also explore the regional variability of gastronomic sociality to gain comparative perspectives. In this way, the proposal of a cultural and territorial pluralisation of gastronomy is supported (Poulain 2005) and is further argued in the succeeding chapter.

Note

1 This chapter was reprinted with permission from Ueda, H. & Poulain, J.P. (2021) What is gastronomy for the French? An empirical study on the representation and eating model in contemporary France. *International Journal of Gastronomy and Food Science*, 25, 100377–100388.

References

Abric, J.C. (1993) Central system, peripheral system: Their functions and roles in the dynamics of social representations. *Papers on Social Representations*, 2(2), 75–78.

Allaire, G. & Boyer, R. (1995) *La Grande Transformation de l'Agriculture : Lectures Conventionnalistes et Régulationnistes*. INRA & Economica.

Andersen, A.H. (2011) Organic food and the plural moralities of food provisioning. *Journal of Rural Studies*, 27(4), 440–450.

Aron, J.P. (1973) *Le Mangeur du XIXe Siècle*. Robert Laffont.

Boltanski, L. & Thévenot, L. (1991) *De la Justification: Les Economies de la Grandeur*. Gallimard.

Bourdieu, P. (1979) *La Distinction: Une Critique Sociale du Judement*. Editions de Minuit.

Chartier, J.F. & Meunier, J.G. (2011) Text mining methods for social representation analysis in large corpora. *Papers of Social Representations*, 20, 37.1–37.47.

Climent-Lopez, E.A., Sanchez-Hernandez, J.L., Canto-Fresno, C., Alonso-Santos, J.L., Ramirez-Garzia, S., Rodero-Gonzalez, V. & Ruiz-Budria, E. (2014) Measuring quality conventions in the food industry: Applications to the wine sector in Spain. *Geoforum*, 56, 148–160.

Corbeau, J.P. (2008) L'indispensable plaisir alimentaire. In J.P. Corbeau (ed.) *Nourrir de Plaisir: Régression, Transgression, Transmission, Régulation?* 5–6. OCHA.

Corbeau, J.P. & Poulain, J.P. (2002) *Penser l'Alimentation: Entre Imaginaire et Rationalité*. Privat.

Csergo, J. (2016) *La Gastronomie est-elle une Marchandise Culturelle comme une Autre? La Gastronomie Française à l'UNESCO: Histoire et Enjeux*. Menu Fretin.

Doise, W., Clémence, A. & Lorenzi-Cioldi, F. (1993) *The Quantitative Analysis of Social Representations*. Harvester Wheatsheaf.

Dupuy, A. (2013) *Plaisirs Alimentaires, Socialisation des Enfants et des Adolescents*. Presses Universitaires de Rennes & Presses Universitaires François-Rabelais.

Fischler, C. (1990) *L'Homnivore*. Odile Jacob.

Fischler, C. & Masson, E. (2008) *Manger : Français, Européens et Américains face à l'Alimentation*. Odile Jacob.

Herpin, N. (1988) Le repas comme institution, compte rendu d'une enquete exploratoire. *Revue française de sociologie*, 29(3), 503–521.

Institut National de la Statistique et de Etudes Economiques (2010) Le temps de l'alimentation en France. https://www.insee.fr/fr/statistiques/1281016 (accessed in January 2019).

Lahlou, S. (1995) *Penser Manger : Les représentations sociales de l'alimentation*. Thèse doctorale. Ecole des Hautes Etudes en Sciences Sociales.

Mathé, T., Beldame, D. & Hebel, P. (2014) Evolution des représentations sociales du bien manger. *Cahier de Recherche*, 316. Crédoc.

Monaco, G.L., Piermattéo, A., Rateau, P. & Tavani, J.L. (2017) Methods for studying the structure of social representations: A critical review and agenda for future research. *Journal of the Theory of Social Behaviour*, 47(3), 306–331.

Moscovici, S. (2001) *Social Representations: Exploitations in Social Psychology*. New York University Press.

Ponte, S. (2016) Convention theory in the Anglophone agro-food literature: Past, present and future. *Journal of Rural Studies*, 44, 12–23.

Poulain, J.P. (2002a) The contemporary diet in France: 'De-structuration' or from commensalism to 'vagabond feeding'. *Appetite*, 39(1), 43–55.

——— (2002b) *Manger Aujourd'hui, Attitudes, Norms et Pratiques*. Privat.

——— (2005) French gastronomie, french gastronomies. In D. Goldstein & Merkele, K. (eds.) *Culinary Cultures of Europe Identity, Diversity, and Dialogue*, 157–170. Editions du Conseil de Europe.

——— (2011) La gastronomisation des cuisines de terroir: sociologie d'un retournement de perspective. In N. Adell & Pourcher, Y. (eds.) *Transmettre, Quel(s) Patrimoine(s)? Autour du Patrimoine Culturel Immatériel*, 239–248. Michel Houdiard.

Rateau, P., Moliner, P., Guimelli, C. & Abric, J.C. (2012) Social representation theory. In P.A. Van Lange, Kruglanski, A.W. & Higgins, E.T. (eds.) *Handbook of Theories of Social Psychology*, 477–497. SAGE Publications.

Reinert, M. (1990) Alceste une méthodologie d'analyse des données textuelles et une application: Aurelia de Gerard de Nerval. *Bulletin of Méthodological Sociology*, 26, 24–54.

Ueda, H. & Poulain, J.P. (2021) What is gastronomy for the French? An empirical study on the representation and eating model in contemporary France. *International Journal of Gastronomy and Food Science*, 25, 100377–100388.

United Nations Educational, Scientific and Cultural Organisation (2010) *Gastronomic Meal of the French* (nomination file, No. 00437).

Warde, A. (2016) *The Practice of Eating*. Polity.

8 Jacques Puisais' Philosophy of Gastronomy and Taste Education

Jacques Puisais (1927–2020) is a philosopher of taste, a gastronome and the inventor of taste education. Despite his undeniable contribution, there exists almost no academic reading of Puisais' philosophy of taste education. Aside from a large volume of studies on its educational impacts (e.g., Reverdy 2011) or institutionalisation (such as Ueda 2018), Puisais has been occasionally mentioned as a prominent example of a proponent of the thematisation of taste (Poulain 2009, p.151) or the pleasure of eating (Corbeau 2008, p.6; Dupuy 2013, p.135) in social sciences and the development of French gastronomy (Pitte 2002, p.166). However, his taste education theory has not yet been fully scrutinised by these social science scholars. Therefore, this chapter aims to fill this knowledge gap in the social sciences of food and taste (Figure 8.1).[1]

8.1 Jacques Puisais: A Man of Two Faces

The approach to understanding Puisais' philosophy is *theoretically* based on the philosophy of taste and the sociology of food, but at the same time *pragmatic* in relation to the practical improvement of taste education currently being promoted both in and outside France. I first discuss data availability, then the pedagogical nature of taste education and the contemporary challenges in its institutionalisation.

Generally, Puisais' professional contributions can be categorised into two fields, namely the culinary and wine industry (Puisais 1981, 1985; Puisais & De Rabaudy 1993) on the one hand and taste education for children on the other hand (Puisais 1999, 2011). Regarding the former's close connection with French gastronomy, Puisais held numerous important positions, such as the founder of the International Wine Academy and as vice-president of the Academy of Gastronomes. However, it is surprising that, when being introduced to other countries (northern Europe and Asia), Puisais was often 'degastronomised' due to the advocates' ignorance of his career as a gastronome or out of fear of the incongruency between the inherent elitism of gastronomy and taste education for all. A previous effort to analyse Puisais' method solely on the basis of the latter literature (Puisais 1999, 2011) highlighted the methodological challenges involved in articulating Puisais' connections with gastronomy (Ueda 2020). Thus, in this chapter,

DOI: 10.4324/9781003341895-11

Figure 8.1 Jacques Puisais (1927–2020)
Source: I thank Nathalie Politzer, educator at Institute of Taste, for providing this photo.

analytical priority is given to the earlier literature on gastronomy (e.g., Puisais 1985). Moreover, to extract the nuanced implications of his text, an additional interview with Puisais was conducted in February 2018.

Since its invention in the 1970s, Puisais' taste education has been increasingly implemented, both within and outside France, whether as the original version or as a variation. Nevertheless, as already seen in Chapter 3, the following common pedagogical characteristics have been identified:

1 Justification: The pedagogies were developed to address children's 'impoverishment of taste' due to the excessive medicalisation, industrialisation and globalisation of food.
2 Objective: Children are aimed at expanding children's capabilities to control their own bodies and construct harmonious relationships with their social environments (including peers).
3 Content: Children learn primarily the physiological mechanism for perceiving taste (comprising five senses), coupled with the relevant socio-cultural dimensions of food.
4 Method: These pedagogies contained (i) theoretical exposition, (ii) tasting food samples and (iii) verbalisation or communication with others, following the three principles of 'respect of subjectivity', 'non-normativity' and 'relativisation of own senses'.

5 Theory: These pedagogies were developed based on the physiology of taste (e.g., Le Magnen, MacLeod) and child psychology (including Chiva 1984).

Although it is certain that Puisais integrated some core elements of the social science of food into his pedagogy, he did not explicitly articulate such imported social theories in his publication. As a consequence of the relative weakness of its socio-cultural dimensions, taste education has been interpreted mainly within the bio-psychological theoretical framework, which turned out to be problematic in the face of the recent scientification and politicisation of taste education, and its distinction from nutritional education (Chapter 5): First, nutritional and sensory scientists have been eager to prove the educational impacts of taste education, such as food neophobia, taste preferences and the ability to discern tastes and smells (Reverdy 2011), which paradoxically resulted in narrowing down the original overarching scope of Puisais' pedagogy. Second, the objective of taste education has often been mistakenly perceived to be merely promoting the consumption of vegetables or traditional recipes in current food policies (e.g., the National Programme of Nutrition and Health and the National Food Programme). Third, taste educators have emphasised their non-normative teaching method to distinguish taste education from existing nutrition education, which is premised on pre-determined norms of 'good food' based on nutritional evidence. This emphasis unexpectedly resulted in the theoretical validity of such non-normativity (discussed later) being challenged. Thus, the socio-cultural dimensions of taste education need to be further considered.

Therefore, the objective of this chapter is to revisit and articulate the philosophical foundation of Puisais' taste education programme in order to rescue it from these contemporary challenges (scientification, politicisation and distinction). To achieve this, I first discuss the transition of the philosophical status of taste to understand 'why taste became an issue of education'. Second, I consider the limits and contributions of recent criticisms of taste education from the perspective of the sociology of taste (Bourdieu). I also tease out some implications of the sociology of eaters (Fischler, Corbeau, Poulain) to advance the socio-cultural dimensions of taste education. Through this philosophical and sociological conceptual framework, three key points of analysis are finally identified, namely beauty, sociability and humanism. I analyse the interrelationships of these elements and clarify what subject (or 'future gastronome') Puisais wished to cultivate during his nearly 50 years of educational endeavours.

8.2 Philosophy of Taste

Taste has been a part of the philosophical debate since Plato touched on the subject, followed much later by Descartes, Lock, Hume, Kant and Brillat-Savarin. I follow Sweeney's (2017) historical analysis to understand how the philosophical status of taste has changed from antiquity to today.

In antiquity, taste was pondered in relation to human spiritual life. In *Phaedo*, Plato considered taste (along with smell and touch) to be hedonistic enough to disturb the human spirit and thus inferior to sight and hearing, which were dissociated from corporeality and functioned as faculties to appreciate music and paintings.

In the Middle Ages, taste was heavily influenced by a religious belief that stigmatised corporeal pleasure. This is traceable to Augustine (354–430), who was afraid that sensory pleasure would lead to the sin of gluttony and recommended fasting to overcome it. During the Middle Ages, Aquinas (1225–1274) confirmed that taste was inferior to the other senses and that this faculty was thus unable to appreciate beauty, which was an object of reason.

In the Renaissance, humanists discovered that food used to be the object allowed to ponder among ancient Greek and Roman thinkers, such as in Plato's *Symposium*. For instance, in his *On Right Pleasure and Good Health*, Platina (1421–481) utilised ancient theories (e.g., Hippocrates) on health to criticise the religious belief that stigmatised sensory pleasure and went on to justify it as contributing to the promotion of health and well-being.

Until the 17th century, taste was considered merely in a literal manner, namely as a physiological function to distinguish foods. However, in the 17th century, taste began to be defined as a critical, mental function to discern the beauty of objects. The condition for this transformation in the philosophical status of taste was prepared by two competing philosophies, Descartes' (1596–1650) rationalism and Lock's (1632–1704) empiricism. Beauty and taste as a discerning function became part of the philosophical debate. The question was whether describing a given object as beautiful was a rational ascription of its actual nature to the perceived object (rationalism) or the pleasant feeling in the subject's experience (empiricism).

This debate continued among the Enlightenment philosophers of the 18th century. Voltaire (1694–1778) similarly distinguished intellectual taste from literal taste and, while both were commonly imminent (not reflective) perceptions, the former was able to be *educated*. Therefore, intellectual taste could be enlightened by reason, while sensual taste was physiologically determined and thus unable to receive such enlightenment. Hume (1711–1776) followed Lockean empiricism, in which beauty was conceived, not as the nature of object itself, as but the experience in the subject's mind. However, he was eventually confronted with a contradiction concerning the subjectivity of taste: If taste is a subjective experience for each subject, no objective beauty can exist. In *Of the Standard of Taste*, he stipulated several conditions for being an authentic critic of beauty, such as having sophisticated taste and being superior in comparative analysis, to overcome this contradiction.

While acknowledging that both literal and critical tastes were based on one's subjective experience, Kant (1724–1804) argued that they differed in terms of the immediacy of perception: Critical taste was a reflective interaction with the object, while literal taste was an immediate pleasant or unpleasant reaction to it. Kant also relied on human 'common sense' to overcome subjectivity in judging the aesthetics. This common sense referred, not to practical knowledge in our everyday language, but to the a priori assumption that there should exist a universal effect (i.e., feeling beautiful) generated by a free play of one's cognitive ability in relation to the object. In other words, Kant theorised that the judgement of beauty could retain its universality because all humans were 'commonly' equipped with these cognitive functions (imagination and understanding). Although Kant certainly advanced the philosophical discussion of aesthetics, the taste of food did not satisfy another condition that he set, namely 'disinterestedness', due to its association with appetite. Therefore, he did not consider it a legitimate function for judging beauty.

This Kantian perspective was gradually overcome in the development of French gastronomy in the 19th century. In *Physiology of Taste* (1825), Brillat-Savarin argued that taste was a legitimate faculty for reflective deliberation and discerning beauty. He integrated the physiological theory of taste, analysed multiple stages of the act of tasting and concluded that taste was not merely an immediate sense (as theorised by Kant), but also the long-lasting act of reflection. He continued that this physiological function was common to all humans and the universality of beauty was thus supported. Since this milestone publication, the philosophical status of taste has gradually improved.

The thesis of universal beauty that originated with Plato was premised on the distinction between beauty and non-beauty. However, the validity of this assumption has been challenged by the postmodernist movement since the latter part of the 20th century, in terms of which each cultural practice is seen to have its own beauty. Puisais' philosophy is not free from this postmodernist influence.

To summarise, taste had long been reduced to being seen as a mere physiological, unreflective and guilty function. However, the thesis of 'taste as an issue of education' was finally proposed after certain intellectual conditions had been satisfied by, for example, the conceptualisations posited during the Enlightenment in the 17th–18th centuries and by the proponents of the emerging French gastronomy and physiology of taste in the 19th century. This philosophical transition provides two lessons. First, the place of 'beauty' in taste education should be examined, since taste has ceased to be seen as a mere physiological function, and its intellectual status is being appreciated but simultaneously challenged in postmodern trends. Second, the nature of 'humanism' in taste education should be clarified because, since Plato, the fundamental motivation for the judgement of beauty has been to prove the human distinction from animals and our unique existence as human beings.

8.3 Sociology of Food

In this section, recent sociological criticisms of taste education in general (both Puisais' programme and other variables) are examined. I argue that the scholars who premised their arguments on sociological theories (such as those of Bourdieu and Foucault) have rightly highlighted the social nature of taste education, but that their criticisms have some limits with respect to the practical improvement of taste education. Given that interpretations of Bourdieu and Foucault tend to be culturally framed (both within and outside France) and that it is difficult to arrive at any general consensus, my intention here is not to resolve all these theoretical issues, but rather to illuminate the complexity of the endeavour to educate one's taste. Lastly, the sociology of eaters is reviewed to extract a more proactive (not critical) sense of the socio-cultural dimensions of Puisais' taste education programme.

8.3.1 *Taste as a Site of Reproduction of Social Inequalities*

The first stream of sociological studies warns us about the risk of social reproduction. In her ethnographic study of taste education practices in the US, Hayes-Conroy (2009) vividly demonstrated that most practitioners, ignorant of the fact that taste is not physiologically determined but socially constructed, were carelessly describing children as having some pre-determined good or bad tastes. Such normative teaching reinforced these children's classed, gendered and racialised identities, and thus reproduced existing social inequalities.

This type of criticism is theoretically based on a series of studies by Bourdieu. In *La Distinction*, he employed the concept of habitus to explain the interrelationships between taste and social classes manifested in various cultural practices. Habitus is theorised as a personal disposition which is structured by the social conditions surrounding a given actor and which, in turn, structures his or her practices. Following the theory of habitus, he defined taste as 'a class culture turned into nature, that is, embodied, which helps to shape the class body' and 'an incorporated principle of classification which governs all forms of incorporation, choosing and modifying everything that the body ingests and digests and assimilates, physiologically and psychologically' (Bourdieu 1979, p.210).

As opposed to the traditional philosophical understanding, for Bourdieu even physiological taste is embodied by class cultures and no longer a given. As exemplified in *La Distinction* in terms of the working-class taste for 'necessity' and the bourgeoisie taste for 'form', this understanding of taste is not limited to the mere preference for certain foods but applied to the total practice of eating, including preparation, presentation and table manners. In this way, (1) the socially-constructed nature of taste and (2) the reproduction of social inequalities are inseparably described in this book. Moreover, because French higher education was harshly criticised in *La Reproduction* (Bourdieu & Passeron 1970), it was *education* that was conceived as a prominent dispositive for reproducing such social inequalities.

Bourdieu's theory of reproduction has had a considerable impact on the view of education since the 1970s, but has often been subjected to misunderstandings. Thus, some specification should be added here so as not to oversimplify its theoretical range. It should not be dismissed that: (1) the socially-constructed nature of taste can be interpreted not only as (2) a danger of reproducing social inequalities but also as (3) a chance to create new social conditions favourable to one's personal development. Education is the very basis of such chances for children to transcend the social inequalities that they face.

Although Bourdieu's view of education was overly-criticised as structurally deterministic, some scholars (Kato 2015; Miyajima 2017) have recently re-evaluated the concept of habitus, arguing that it was originally conceived as a flexible dispositive, rather than as a deterministic structure, for rightly explaining one's conditional freedom and agency, and that Bourdieu proactively positioned education as an opportunity to grant children 'intellectual technologies' (e.g., critical thinking, collecting data, developing figures and tables) that would enable them to disclose and overcome social inequalities. Thus, to rescue Bourdieu's original intention, education should rather be conceived as both (1) reproductive and (2) creative.

In sum, these sociological studies warn us about the potential danger that 'inappropriate' practices (such as the careless prescription of good taste) would lead to reproducing children's social inequalities. Nevertheless, a more proactive significance of taste education should be explored, rather than merely being obsessed with the thesis of social reproduction.

8.3.2 *Docile or Dominated Tongues*

The second type of criticism is constructed in accord with the following logic (such as that of Leer & Wistoft 2018): The current taste education is not a learning experience for children that enables them to acquire free and respected sensory experiences, but rather a form of control, prescribing for them the 'correct' tastes based on nutritional knowledge. Although two sub-groups of criticism – nutritionism and docile tongues – can be identified in this logic, I focus on the latter and its underlying theory, that is, the Foucauldian theory of power.

In his mid-term work, *Surveiller et Punir*, Foucault (1975) proposed the concept of disciplinary power by questioning 'how subjects were produced', which was very different from the previous thematisation of 'who exercised power over whom'. To take the example of a panopticon, it instils prisoners with a sense of constant surveillance, which causes them to discipline themselves, instead of being governed by a forced, violent and external power. This disciplinary technology produces docile bodies in the subjects, who should correspond with certain norms, and it has been applied to various modern institutions, including schools. At the beginning, Foucault emphasised its negative implication and, in such a sense, Leer and Wistoft (2018) also criticised the docile tongues produced in taste education. However paradoxically, as Foucault argued, none can escape this disciplinary power. Thus, it seems that the use of Foucauldian theory in the criticism of taste education ended up being trapped by the impossibility of avoiding this negative scenario.

The solution had already been proposed in Foucault's later work (1984a, 1984b). The 'power relations' (including disciplinary power) were distinguished from the absolute obedience of 'domination'. The former was then positively redefined as containing the possibility for constant resistance and criticism (Seki 2001). If one follows this argument to its logical conclusion in relation to the polemics of taste education, it would be appropriate to rephrase Leer and Wistoft's (2018) wording by saying that the target of their criticism is not docile but dominated tongues. Two implications can be obtained from the thesis of dominated tastes.

First, theoretically speaking, Foucauldian domination cannot be attributed to taste education that already includes the principles of non-normativity and respecting one's subjective sense. Thus, as is the case with social reproduction, this criticism is applicable only to 'inappropriate' practices. Second, even if dominated situations were produced, Foucault (1984a) argued that one could escape its effects by having '*souci de soi*' (care of the self), that is, shifting one's focus from the external to the internal norms that one defines on one's own and reconstructing them as the subject of a free act. This strategy is not self-contained, but rather requires others as inter-subjective media to facilitate one being reconstructed as the subject.

Although Foucault (1984b) originally proposed *souci de soi* in relation to a subject with a good command of sexual pleasure, commonalities can be identified with taste education, the objective of which is to expand one's capabilities to control the 'self' and construct harmonious relationships with 'others'. It can be said that the 'appropriate' practice of taste education can eventually lead the subjects' tasting capabilities out of such domination.

8.3.3 *Limits of Critical Attitudes: From Good Food to Good Subject (Eater)*

In relation to Bourdieusian and Foucauldian criticisms, the place of 'norms' in taste education came to be seen as a debatable issue. In these critical studies, the relationships between power and knowledge were analysed by employing methods to diagnose how norms were produced in intertwined power relations and to challenge the pseudo-transcendency of such norms. In fact, these methods had already been applied to food education to disclose the unequal power distribution in constructing what should be educated (e.g., Takeda 2008; Kimura 2011).

However, the final resolutions of these previous studies converged only into having 'critical attitudes' towards any definition of 'good' food. This implication would be reasonable for a sociological analysis, but does not seem helpful for educational purposes. Hayes-Conroy (2009) a little more daringly proposed certain concrete forms of criticism (e.g., directed at expert knowledge and dualism between subject and object). Nevertheless, a single focus on critical attitudes does not explain the whole educational expectation and makes it impossible to construct any educational norm regarding what subject should be pursued in schools. This line of counter-argument is not entirely new, because there is an ongoing debate on how modern education could reconcile with postmodernist critics (Geshi 2016).

To summarise, a series of studies on the philosophy of taste upgraded its status from the merely physiological to the intellectual, but its social nature was not yet examined. The sociology of taste then advanced this aspect and highlighted the potential danger of inappropriate practices, such as reproducing children's social inequalities and producing dominated (unfree and uniformed) tongues. Again, this article is not intended to resolve the sociological debate on the place of norms in education. Nevertheless, even to advance such debate, exploration should not stop with the deconstruction of any norm of 'good food' but go on to construct some clear norms regarding 'good subject (eater)'.[2] To better situate this position, a new theoretical tradition of the sociology of eaters is reviewed below.

8.3.4 *Taste and Sociability in the Sociology of Eaters*

It is widely known that Puisais enjoyed close personal and scientific interactions with pioneers of the sociology of eaters (such as Fischler, Corbeau and Poulain) in the institutionalisation of taste education. This academic tradition originally emerged in the 1970s with a mission to go beyond the ambiguity of Durkheim's (1895) definition of the social fact and the relative neglect of subjectivity in the sociology of taste (Bourdieu 1979). Instead, the sociology of eaters is clearly positioned within the Maussian (1935) tradition and Morin's (1973) transdisciplinary approach. It is aimed at exploring the connections between physiology, psychology and sociology that operate with regard to the act of eating (Poulain 2002, p.173). Sociologists of eaters have contributed to articulating the bio-psycho-socio-cultural dimensions of food-related objects, which were previously reduced to single disciplines, such as taste and gastronomy (Poulain 1985, 2017), gastro-anomy (Fischler 1990), sociability (Corbeau 1997), obesity (Poulain 2009) and pleasure (Dupuy 2013).

For further investigation of Puisais' taste education, I highlight three of the most relevant discussions in the field of the sociology of eaters. First, taste is positioned as a 'total' entity at the bio-psycho-socio-cultural interface by going beyond the previous positionings, whether merely physiological, philosophical or social. I will articulate the similar thematisation in Puisais' philosophy below. Second, the concept of sociability is positioned as a driver of social change. Sociability refers to an interactive process in which individuals either decide to accept the social role defined by their sociality, or to develop creative dynamics through the interactions they seek to bring about (Corbeau 1997, p.15). This account of creative possibilities of social change can partly support the counter-argument against normless education (as described above). Third, the recent work on the pleasure of eating (Dupuy 2013) guides us to distance ourselves from the traditional theories, whether as an inquiry into some pre-determined 'good command' of pleasure or as a physiological driver of healthy behaviours, and focus more on the subjectivity of sensory experiences. All of these are kept in mind when exploring Puisais' thought.

8.4 The Philosophical Foundation of 'Future Gastronomes'

8.4.1 *Beauty: The Totality of Food*

In the following sections, the focus is on three key elements – beauty, sociability and humanism – to explore what subject – the future gastronome (Puisais 1999, p.8) – Puisais wished to cultivate in taste education. In so doing, I also discuss its conceptual connection with French gastronomy and the thought of the very first philosopher and educator of taste, Brillat-Savarin, whose contributions were identified in the proceeding section. In common with the thinkers after Brillat-Savarin, Puisais also clearly situated taste as the function of appreciating the beauty of the table.

> Taste education is an issue of time and humbleness. The one does not discover opera with Wagner's Parsifal, or painting with Mondrian, or literature with Proust [...] It will be the same with high gastronomy and it is not synonymous with complication. Taste for the table cultivates itself, as is the case with the other arts (Puisais 1985, p.33).

Regarding this text, it is worth taking note of three characteristics. First, gastronomy is positioned as a legitimate 'art', along with the other arts (e.g., music, painting). Second, there is therefore no hierarchy of senses, in contrast to the traditional view of taste being inferior to hearing and sight. Third, taste is to be cultivated over time for the one to appreciate beauty in the art of the table (gastronomy); and this is where taste education becomes necessary. To explain the meaning of such beauty at the table, Puisais defines the three properties of 'food and drink [...] to play gastronomy' (2011, p.21).

1. Bien – the nutritional properties for the reinforcement of bodies
2. Bon – the hygienic properties for the health of bodies
3. Beau – the organoleptic (or gustative) properties for the pleasure of bodies and 'minds'

Subsequently, the nutritional and hygienic qualities of food are categorised as the 'first quality' to be quantitatively measurable, while the gustative quality is valued as the 'second quality' to be perceived by the five senses and as the object of judgement of beauty. It would be appropriate to consider that this second quality refers, not only to the organoleptic and sensorial, but also the affective and symbolic aspects of food, because the senses are to be perceived as being in a constant interaction with one's emotions and memories (Puisais 2011, pp.26–30). It is this focus on the 'total' qualities of food that distinguishes taste education from nutritional education, which is focused primarily on the first quality of food. The similarity should also be highlighted between Puisais' and the sociologies of eaters in thematising food or taste as bio-psycho-socio-cultural.

Moreover, it should be recalled that, since Plato, the fundamental motivation for the judgement of beauty has been to prove the unique existence of human beings, which also resonates with Puisais' distinction between *nourriture* and *aliment*.

> Puisais' definition of 'taste'
>
> It [taste] is the emotional totality that fulfils one's mind and memory, and it is where there is a difference between *nourriture* [foodstuffs] and *aliment* [total food]; in English there is only one word, 'food', to translate this [difference], while in French we have [both] *nourriture* that nourishes the body and *aliment* that [also] satisfies the spirit.
>
> (Hervier 2009, pp.56–57)

In sum, the beauty in Puisais' taste education means *aliment*, the totality of food and eating, which is in harmony with the sociology of eaters. This objective is not to deny the importance of nourriture, the nutritional and hygienic qualities of food, but rather to play the humanistic 'game of the spirit' (Puisais 2011, p.27) at the bio-psycho-socio-cultural intersection. Nevertheless, he strategically prioritised the aesthetic quality of food over the nutritional in the face of the increasing medicalisation and industrialisation of food that has occurred since the 1970s.

8.4.2 Sociability at the Table

The second element is sociability, which has its origins in the salon culture of post-revolutionary France. To see how Puisais appropriated this concept, I first discuss the philosophy of Brillat-Savarin. As reviewed in the proceeding section, in his anatomical analysis of taste, Brillat-Savarin identified its reflective nature. The following texts indicate his underlying motivation for such a theorisation of taste (Brillat-Savarin 1825[1982]):

> [...] the pleasure of the table should be well distinguished from the pleasure of eating. The pleasure of eating is the real and direct sensation of satisfying one's need, while the latter is the reflective sensation that is born from various situations of facts, places, things and personages that are accompanied at the table [...] The pleasure of the table is unique to the human species; it supposes preliminary cares for the preparation of meals, the choice of the places and the invitation of guests.
>
> (pp.170–171)

> The effect of the gourmandise on the sociability –
>
> The gourmandise is one of the principal bonds of the salon; it is this [gourmandise] that gradually enlarges the spirit of conviviality that always unites people of differing status, merges them into a single whole, facilitates the conversation and alleviates conventional inequalities.
>
> (p.147)

Largely influenced by an idealising effect of post-revolutionary egalitarianism, Brillat-Savarin had no doubt that such pleasure at the table, or conviviality (sociability at the table), could be realised by all the people, regardless of social status. However, it is also true that the post-revolutionary salon was in reality open exclusively to the upper social class. In such an elite world, the reflective taste that he defined was to serve as an objective descriptor of social status. It was therefore necessary to 'educate taste' to enjoy such sociability (Hashimoto 2019). Keeping this inherent challenge in mind, one should ask how Puisais dealt with the elitism associated with sociability in his theorisation.

Puisais used the two terms 'sociability' and 'conviviality' interchangeably. Sociologically, sociability is defined as a playful association in which an individual's pleasure is contingent on the joy of others (Simmel 1949), while conviviality is defined as a playful interaction in which those who eat together are motivated by a collective desire for amicability and cordiality (Phull et al. 2015). In other words, conviviality refers to the realisation of sociability at the table; thus, the two concepts are similar to each other. As exemplified in the following excerpt, such a conception was also true of Puisais (1981):

> It [conviviality] enables the exchanges between humans who favor the idea of tolerance. Or, can we find a melting pot which is more universal and fraternal than the 'taste'? [...] Being united around the same table [...] will be able to resolve misunderstanding and hatred, consolidate friendship, bring concepts closer and make the impossible possible. Through such taste, conviviality becomes an art that helps to promote popular and authentic cultures, which allows for liberating and soliciting a general reflection on human behavior and flourishing in one's existence [as human] and the art of living.
>
> (p.155)

On the first reading, Puisais' concept of conviviality seems almost the same as that of Brillat-Savarin's sociability. This text certainly implies his humanist philosophy but does not tell us much about how he reacted to the inherent elitism. To investigate this, I first move to the discussion of his humanism and later come back to his norm regarding the 'good' subject.

8.4.3 *Humanism: Paideia and Philanthrōpia*

The two key elements, beauty and sociability, were positioned as drivers for promoting 'the game of the spirit' and 'flourishing in one's existence as humans'. This is where humanism, the third element in Puisais' philosophy, is of vital importance in terms of connecting the first two elements.

Humanism is an ambiguous concept. For example, Renaissance humanism (e.g., that of Erasmus, Rabelais) was a protest against the corruption of the Catholic Church in the Middle Ages, whereas German humanism (such as that of Goethe) was a polemic against the rationalist dilution of impulsive, affective and emotional integrity, whereas Marxist humanism was a polemic against capitalism. In the end, it refers to all forms of 'protests against self-alienation' which are 'dependent on political, economic and religious contexts' (Mutai et al. 1956).

The point here is what form of humanism Puisais wished to achieve. To elucidate this, it would be useful to operationally distinguish two meanings of humanism, or *humanitas* in its original Latin form (Shoji 1968; Nejime 1984; Ishigami 2010). The first meaning is *paideia*, the culture which can be pursued only by the humans. The second one is *philanthrōpia*, fraternity with others. These two meanings of *humanitas* were the qualities of a 'good *polites* (citizen)' and thus not distinct for the ancient Greeks and Romans, including Cicero, who coined this term (Ishigami 2010). However, these two meanings are distinguished today, so the temporal distinction of *paideia* and *philanthrōpia* shall be retained when analysing the humanism Puisais wished to achieve in his taste education. The key to understanding Puisais' concept of humanism lies in his understanding of gastronomy, for which Brillat-Savarin (1825[1982]) gave the first systematic definition:

> Gastronomy is the reasoned comprehension of everything that is related to man as the one who nourishes himself [...] gastronomy is attached to: natural history for the classification of food substances; physics for the examination of their compositions and qualities; chemistry for the diverse analyses and decompositions; cooking for the art of preparing the dishes and making them pleasant to the taste; commerce for the research into the means of buying what is consumed in the best possible market and selling what is available to sell the most advantageously; lastly, political economy for the resources to which the tax is introduced and for the means of exchange that are established between the nations.
>
> (p.62)

Brillat-Savarin enjoyed diverse professional experiences, from as advocate to diplomat, politician and musician. This life-course as a 'universal operator of discourse' eventually produced a broad sense of gastronomy, extending from natural history to political economy. As Barthes (1989) called him as an encyclopedic 'humanist', this definition of gastronomy can be formulated as a *paideia*-type humanism in the vocabulary of this analysis.

As noted in the introduction, Puisais likewise served as an authentic gastronome in his career. Thus, it is not difficult to imagine how much his capabilities to appreciate the total quality of food and wine, and to socialise with others at the table were requested, the success of which eventually led to his current recognition as a 'poet' (Polony 2019), 'philosopher' (Cody 2009) and 'humanist' (Corbeau 2008). The most symbolic statement that he was a humanist can be seen on the statue of the French Institute of Taste that he established in 1976:

> Objective and Method –
>
> The French Institute of Taste is interested in the history, anthropology and biology of food.
>
> The Message of Taste –
>
> The French Institute of Taste was established to continue as a center of reflection and derive answers for this theme [of reflection...] a gastronomic language expresses itself as the authentic carrier of humanism.

Puisais' humanism as *paideia* was observed, not only in this statement, but also in the actual organisation of the scientific committee, composed of French historians, sociologists and food psychologists, and annual conferences themed on the totality of taste and food (as seen in Chapter 5). Moreover, he wished that this humanism would be applied to both professionals and children:

> The child who truly tastes a carrot, an apple, a glass of milk, knows where they come from and how they are produced, and, at the same time, the pleasure; [for this] he learns history, geography, and a great deal of humanity
> (interview with Puisais, cited from Berger 2013).

It is worth recalling the concept of *aliment* (totality of food), as discussed in the previous section. Puisais defined the humanism in gastronomy elsewhere as 'the vertical and horizontal [total] relations of humans, materials and environments' (Hervier 2009, pp.63–64). Thus, *paideia* can thus be equated with *aliment*, which is one of the educational objectives in taste education.

Interestingly, however, his humanism was not limited to this total understanding of food. As one of the practitioners closest to Puisais (Politzer 2016) reflected, he 'developed the taste education method based on [...] the spirit of humanism, as completely opposed to elitism' (p.4). Puisais also embraced the spirit of *philanthrōpia*, a fraternity characterised by respect for others regardless of religion, social class or ethnicity. This dimension of humanism was clearly stated in his interview (P: Puisais, A: author):

A: Could you tell us your thoughts about humanism in taste education?

P: One has to read Rabelais. Behind his book, there was a philosophy of laicité, in which having which type of religions was not questioned. It is the philosophy of respecting every individual's unique 'moment'. To take an example, there are 150 winemakers in Chinon. There are Catholic and Protestant producers, and rich and poor ones. But these origins are never questioned, because the truth that they cultivate the same terroir of Chinon never changes. Humanism is like that. It is the philosophy of respecting all humans.

It is no surprise that Puisais sympathised with Rabelais, because both thinkers were born in Chinon. Rabelais' Gargantua and Pantagruel are more generally appraised as the embodiment of a paideia-type humanism, the acquirement of the ancient Greek and Roman cultures to oppose the corruption of Catholic theology in the Renaissance (Watanabe 2005). However, Puisais interpreted it differently, as a philanthrōpia-type humanism. Understood in this way, one can easily identify the similarity of this *philanthrōpia* with the sociability or conviviality in Puisais' taste education, because both are directed at building respectful relationships with others.

8.4.4 *Le goût juste* – Normative Approach in Taste Education

It was so far confirmed that the subject to be pursued in taste education is a humanist eater who has the two inseparable capabilities of *aliment* (*paideia*) and sociability (*philanthrōpia*). However, the question of how Puisais reacted to the elitist normativity inherent in taste education remains to be answered. Therefore, I address this point by analysing Puisais' concepts of gastronomy, good food and what he calls the '*goût juste*' (just taste). This question was put to Puisais in an interview.

A: What does gastronomy mean for you?

P: Gastronomy is the art of doing good food (*faire bonne chère*). It is knowing how to drink and eat. To know how to eat is to know how to taste. Without knowing how to taste, one cannot drink or eat. For me, to swallow without tasting is only to ruin the palate (*avaler sans goûter, n'est que ruine du palais*) […]

A: Is this gastronomy practiced only at high-class restaurants, as often said?

P: Gastronomy is my breakfast. Everyone can do this with a slice of ham and bread with confiture. We should stop saying that gastronomy is something luxurious.

Here, Puisais consciously criticised the elitist nature of gastronomy. Moreover, his criticism can be further identified in the following argument.

A: You already said that, but what does good food (*bonne chère*) mean for you?

P: It is knowing how to drink and eat. Gastronomy looks like the balance between the dry and the humid […] as written in my book, *Le Goût Juste*, all the wines have their balanced pairs with dishes.

A: Then, what about the meaning of good product (*bon produit*) for you?

P: Food cannot be good or bad. It is only 'just' or not. It is whether the food is appropriate for climates, lands, and producers.

These definitions of 'good food' indicate the following logic: Food cannot be good or bad in itself; it is, rather, a neutral object. If the proper balance with its social environment is achieved, that neutral food can be enhanced to the level of gastronomy. Puisais harshly criticised the elitist definition of gastronomy as an assemblage of luxurious foods or dishes. For him, the *goût juste* is not the concept of 'good food' but the concept of 'good subject (eater)' who can keep aiming for such a fair balance of given foods and situations.

But what about the elitism inherent in taste education? Certainly, if we follow the previous criticisms (Bourdieu, Foucault), the 'knowing how to taste' itself should be considered as already elitist, for it requires a rich food experience, sophisticated taste and a high level of sociability, and excludes from the gastronomic realm those who 'do not know how to taste'.

However, as far as education is concerned, this line of reasoning seems a little too narrow. To recall the previous discussion on the norms in education, (1) the effect of education should not be reduced merely to the reproduction of social inequalities, and (2) certain 'norms' should be proposed for it to be named education. If the intention of this paper is accepted, I move on to argue for a more proactive sense of the goût juste.

> There exists no rule. No theorem of taste.
> In contrast, we can talk about privileged pairings
>
> (Puisais 1985, p.70).

This absence of pre-determined rules of the *goût juste* is consistent with the non-normative teaching method of taste education. Certainly, the 'privileged pairings' might carry the risk of being mistakenly taken as elitist. However, Puisais (1985, p.8) was merely 'suggesting and opening the imagination of others [...] without any claimed imposition' of the *goût juste* that he himself preferred. Moreover, for Puisais (2008) there is no superiority of the tastes of one person over that of another, or of the tastes of one moment over those of another moment within the same person:

What is the goût juste?

- Two identical individuals do not exist.
- The same individual never relives the same moment.
- It is the lived memory that prepares the experience of the moment.

What he wished to achieve in taste education was to enable every child to reflect on her existentially unique and subjective taste, and to keep seeking the pleasant balance, which she believes, of a given food with a given situation. This is what Puisais (1999, p.10) defines as the 'development of personality' that applies to every child, regardless of their origins. Here, Puisais is no longer preoccupied with the notion of universal beauty (or good) in the traditional philosophy of taste. Rather, he advocates for the existential uniqueness of every single individual.

In sum, this section was aimed at addressing two questions. The first was whether Puisais' taste education was normless, or, if any, what norm was to be pursued. It was found that Puisais clearly embraced an educational norm; however, this did not refer to 'good food' but rather 'good eaters (subjects)'. Indeed, Puisais wished to cultivate humanist eaters with the capabilities for *aliment* (*paideia*) and sociability (*philanthrōpia*). The second question was whether his approach to constructing these norms could be considered elitist. As indicated in the previous section, it depends on how appropriately the actual educational practice is performed. However, one should not ignore the fact that Puisais was critically conscious of this inherent elitism but, at the same time, aimed to situate the normativity of the subject more proactively for the development of personality. His emphasis on the free, creative and subjective sensory experiences was ultimately consistent with the claims in the recent scholarship on the sociology of sociability (Corbeau 1997) and the pleasure of eating (Dupuy 2013).

8.5 Concluding Remarks

The objective of this chapter was to elucidate the philosophical foundation of Puisais' taste education. To achieve this, I first discussed the transition of the philosophical status of taste and clarified that the theme of 'taste as an issue of education' was initially proposed after satisfying certain modern philosophical conditions (the Enlightenment of the 17th–18th century, French gastronomy and the first emergence of physiology of taste in the 19th century); and it was finally crystalised into Puisais' taste education programme, with the help of new sciences of taste in psychology, neuroscience and physiology (e.g., Chiva 1984; Le Magnen 2001), as a protesting reaction to the self-alienation caused by the increasing medicalisation, industrialisation and globalisation of food since the 1970s.

Second, I scrutinised recent criticisms of taste education found in sociological studies and concluded that they helped to highlight the socially-constructed nature of taste and the potential risk of inappropriate practices, such as the social reproduction and domination of free and diverse sensory experiences. However, a more proactive and constructive approach to Puisais' philosophy was requested. Thus, the theoretical contributions of the sociology of eaters were visited to extract useful implications on taste and sociability. Third, using the relevant literature and the interview with Puisais, three key points, namely beauty, sociability and humanism, were scrutinised and the concrete figure of a 'future gastronome', a subject to be pursued in taste education, was elucidated (Figure 8.2).

This subject was composed of two capabilities: (1) Appreciating *aliment*, the total (bio-psycho-socio-cultural) values of food and (2) playing sociabilité/convivialité, shared pleasure with others at the table. In taste education, every individual is encouraged to go beyond pre-determined good tastes and seek for the *goût juste* in which he or she believes, to eat more humanistically.

Although the significance of Puisais' philosophy might be approached differently in future studies, one possible contribution of this exploration was to concretise the characteristics of this humanist-gastronome as the educational norm that Puisais wished to achieve. This finding would also be pragmatically effective to reorientate the institutionalisation of taste education within and outside

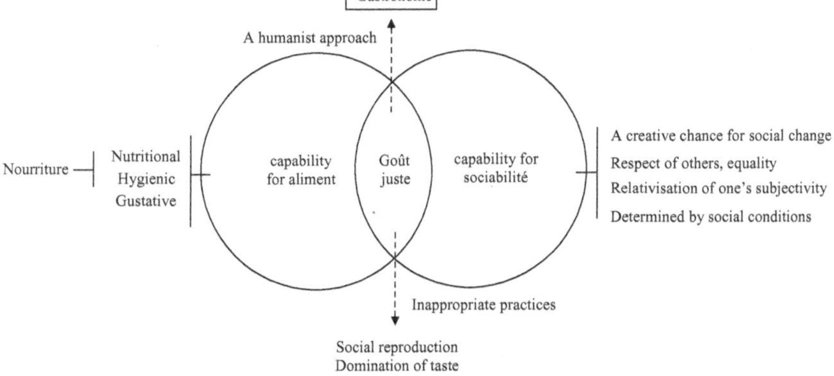

Figure 8.2 The Future Gastronome as the Norm of Puisais' Taste Education
Source: Reprinted with permission from Ueda (2021).

France, given the contemporary challenges (scientification, politicisation and distinction) that have obscured the fundamental assumptions of Puisais' philosophy.

Notes

1 This chapter was reprinted with permission from Ueda (2021) The philosophy of taste education: Regarding Jacques Puisais as a contemporary humanist-gastronome. *International Journal of Gastronomy and Food Science*, 25, 100385–100394.

2 Indeed, without any norm of what subjects should be pursued in education, a 'critical attitude' would lose its coherence and end up with opportunist attacks on given norms.

References

Aron, J.P. (1973) *Le Mangeur du XIXe Siècle*. Robert Laffont.

Barthes, R. (1975) Lecture de Brillat-savarin. In J.A. Brillat-Savarin (ed.). *Physiologie du Goût*, 7–33. Hermann. Translated by R. Howard (1989) *The Rustle of Language*, 250–270. University of California Press.

Berger, D.J. (2013) Jacques Puisais anime Musica vini. Article on August 18th on the website of Mtonvin.net. (accessed in January 2019).

Bourdieu, P. (1979) *La Distinction: Une Critique Sociale du Judement*. Editions de Minuit.

Bourdieu, P. & Passeron J.C. (1970) *La Reproduction : Elements d'une Théorie du Système d'Enseignement*. Editions de Minuit.

Brillat-Savarin, J.A. (1825) *Physiologie du goût*. Edited by J.F. Revel (1982) Flammarion.

Chiva, M. (1984) *Le Doux et l'Amer*. Presses Universitaires de France.

Cody, E. (2009). Philosopher of taste. Article on November 7th on the Washington Post. https://www.washingtonpost.com/wp-dyn/content/article/2009/11/06/AR2009110604194.html (accessed in January 2019).

Corbeau, J.P. (1997) Socialité, sociabilité et sauce toujours. In J. Duvignaud, & Khaznadar, C. (eds.), *Cultures, Nourritures*, 69–81. Babel-Actes Sud.

——— (2008) L'indispensable plaisir alimentaire. In J.P. Corbeau (ed.) *Nourrir de Plaisir: Régression, Transgression, Transmission, Régulation?* 5–6. OCHA.

Dupuy, A. (2013) *Plaisirs Alimentaires, Socialisation des Enfants et des Adolescents*. Presses Universitaires de Rennes & Presses Universitaires François-Rabelais.

Fischler, C. (1990) *L'Homnivore*. Odile Jacob.

Foucault, M. (1975) *Surveiller et Punir : Naissance de la Prison*. Gallimard.

——— (1984a) *Histoire de la Sexualité (vol.3) : Le Souci de Soi*. Gallimard.

——— (1984b) *Histoire de la Sexualité (vol. 2) : l'Usage des Plaisirs*. Gallimard.

Geshi, A. (2016) *Kyoiku shiso no posuto modan [The Postmodern Educational Philosophy: Beyond the Post-war Educational Sciences]*. Keiso Shobo.

Hashimoto, C. (2019) *La Naissance du Gourmand : Grimod de la Reynière et la Révolution Française*. Presses Universitaires François-Rabelais.

Hayes-Conroy, J. (2009) *Visceral Reactions: Alternative Food and Social Difference in Two North American Schools*. Pennsylvania State University, Ph.D. Thesis.

Hervier, D. (2009) *Lettres de Goût: L'Abécédaire de Jacques Puisais*. Editions Féret.

Ishigami, Y. (2010) Humanism nit suite [On humanism]. *Review of the Department of Correspondence Education of Soka University*, 13, 108–124.

Kato, H. (2015) *Bourdieu: Tatakau Choshikijin [Bourdieu: The Intellectual in Battle]*. Kodansha.

Kimura, A.H. (2011) Food education as food literacy: privatized and gendered food knowledge in contemporary Japan. *Agriculture and Human Values*, 28(4), 465–482.

Leer, J. & Wistoft, K. (2018) Taste in food education: A critical review essay. *Food and Foodways*, 26(1), 1–21.

Le Magnen, J. (2001) My scientific life: 40 years at the College de France. *Neuroscience and Biobehavioral Reviews*, 25(5), 375–394.

Mauss, M. (1968) Les Techniques du Corps. In M. Mauss (ed.) *Sociologie et Anthropologie*. Presses Universitaires de France. Translated by T. Aji & Yamaguchi, T. (1976) *Sociology and Anthropology II*, 121–156, Kubundo.

Miyajima, T. (2017) *Bunkateki Saisesan no Shakaigaku [The Sociology of Cultural Reproduction (revised edition)]*. Fujiwara Shoten.

Morin, E. (1973) *Le Paradigme Perdu : La nature humaine*. Seuil.

Mutai, R., Taniagawa, T. & Uehara, S. (1956) *Humanism no Shiteki Tenkai [The Historical Development of Humanism]*. Hobunkan Shuppan.

Nejime, K. (1984) Alcune considederazioni sull'umanesimo del Rinascimento italiano. *Studi Italici*, 33, 105–128.

Phull, S., Wills, W. & Dickinson, A. (2015) Is it a pleasure to eat together? Theoretical reflections on conviviality and the Mediterranean diet. *Sociology Compass*, 9(11), 977–986.

Pitte, J.R. (2002) *French Gastronomy: The History and Geography of a Passion*. Columbia University Press.

Politzer, N. (2011) Où en est l'éducation au goût des jeunes en 2011. Dossier Interne. Translated by I. Tajiri. (2016) *Mikaku kyoiku no tanjyo to hensen* [The birth and transition of taste education]. In K. Ishii, Puisais, J., Sakai, N. & Tajiri, I. (eds.) *Kodomo no tameno Mikaku Kyoiku [Taste Education for Children]*, 2–13. Kodansha.

Polony, N. (2019) Jacques Puisais ou le goût de la France. Marianne, 1168.

Poulain, J.P. (1985) *Anthropo-sociologie de la Cuisine et des Manières de Table*. Thèse pour le Doctorat de Sociologie, Universitaire de Paris VII.

——— (2002) *Manger Aujourd'hui, Attitudes, Norms et Pratiques*. Privat.

——— (2009) *Sociologie de l'Obésité*. Presses Universitaires de France.

——— (2011) La gastronomisation des cuisines de terroir: sociologie d'un retournement de perspective. In N. Adell & Pourcher, Y. (eds.) *Transmettre, Quel(s) Patrimoine(s)? Autour du Patrimoine Culturel Immatériel*, 239–248. Michel Houdiard.

——— (2017) *The Sociology of Food: Eating and the Place of Food in Society*. Bloomsbury.

Puisais, J. (1981) *Le vin se met à table*. Éditions Marcel Valtat.

——— (1985) *Le Goût Juste des Vins et des Plats*. Flammarion.

——— (1999) *Le Goût chez l'Enfant : l'Apprentissage en Famille*. Flammarion.

——— (2011) Et Si Nous Refusions la MacDonaldization du Goût! Délicéo.

Puisais, J. & De Rabaudy, N. (1993) *Le Goûteur et le Voluptueux: Scènes de la Vie Gourmande*. Éditions Julien.

Puisais, J., MacLeod, P. & Politzer, N. (2002) *Le Goût et les 5 Sens*. Cédérom Odile Jacob Multimédia-Screen.

Puisais, J. & Zékri-Hustel, R. (2010) *Le Temps du Goût*. Privat.

Reverdy, C. (2011) Sensory education: French perspectives. In V.R. Preedy, Watson, R.R. & Martin, C.R. (eds.) *Handbook of Behavior, Food and Nutrition*, 143–157. Springer.

Seki, Y. (2001) *Foucault no Kenryokuron to Jiyuron [Foucault's Theories of Power and Freedom]*. Keiso Shobo.

Shoji, M. (1968) *Humanism no Kyoiku Shiso [The Educational Philosophy of Humanism]*. Toko Shoin.

Simmel, G. (1949) The sociology of sociability. *American Journal of Sociology*, 55(3), 254–261.

Sweeney, K.W. (2017) *The Aesthetics of Food: The Philosophical Debate about What We Eat and Drink*. Rawman & Littlefields.

Takeda, H. (2008) Delicious food in a beautiful country: Nationhood and nationalism in discourses on food in contemporary Japan. *Studies in Ethnicity and Nationalism*, 8(1), 5–30.

Ueda, H. (2018) A comparative analysis of promotion systems for taste education in Japan, France, and Italy. *Journal of Food System Research*, 25(2), 48–70.

——— (2020) History and pedagogical nature of taste education in Japan, France and Italy through textbook analysis. *Journal of Food System Research*, 27(2), 48–66.

——— (2021) The philosophy of taste education: Regarding Jacques Puisais as a contemporary humanist-gastronome. *International Journal of Gastronomy and Food Science*, 25, 100385–100394.

Watanabe, K. (2005) *France Humanism no Seiritsu [The Birth of French Humanism]*. Iwanami Shoten.

9 Sociology of Japanese Gastronomy

A Comparative Perspective

Thus far, I have argued that it is essential to understand the cultural tenets of taste education to fully grasp its theoretical and practical development. The gastronomic foundation of French taste education was extensively explored in the preceding chapters. Now, I turn to Japanese gastronomy, but we immediately confront a challenge in setting such an analytical focus. In Japan, gastronomy has often been mistakenly interpreted as equivalent to hedonism, elitism or gluttony, with some guilty connotations (see Chapter 1). In contrast to the French view of gastronomy, it cannot be immediately considered the 'common property' of the Japanese population.

Then, what about *washoku*, which is celebrated as an intangible cultural heritage of all Japanese? A prevailing confusion about its historical development, both among experts and lay people, is problematic (see Chapter 6). Therefore, I propose 'Japanese gastronomy' as a new analytical concept that entails the twofold structure of (i) the genesis of *kaiseki* (*haute cuisine*) prior to modernisation and (ii) its subsequent transmission to contemporary food cultures (*nihon-shoku*). This classification will make it possible to maintain comparative perspective with French gastronomy. Note that, due to this operational conceptualisation, popular traditions prior to modernisation (i.e., before the Meiji period) are not explored here. Thus, Japanese gastronomy is not a synonym for 'Japanese food cultures' in general terms.

Hereinafter, I adopt Poulain's (2017) sociological approach, as it has been applied to French gastronomy, to analyse the philosophical, social or cultural and religious conditions for developing Japanese gastronomy. I also cover two themes that are not addressed in UNESCO's heritisation project, namely the view of nature and the healthiness of Japanese gastronomy. The main focus of the following discussion is to propose a basic research draft of Japanese gastronomy, not an in-depth historical analysis. The aim is to help create common ground for an international comparison of gastronomy and, to a certain extent, mitigate the *sakoku* (isolation) in Japanese food culture studies.

DOI: 10.4324/9781003341895-12

9.1 Food Culture in the Philosophy of the 'Unity of Subject and Object'

9.1.1 *Japanese Philosophy of the 'Unity of Subject and Object'*

To recall the case of French gastronomy, a philosophical setback was established through dialectical dynamism, whereby the status of the connection between food and the mind shifted from being considered undivided (from antiquity to the Renaissance) to completely divided (Cartesian) and then as a total social fact (1960s and onwards). Yet, how can this thesis be extended to regions beyond the immediate influence of Descartes, including Japan? Cartesian philosophy was surely imported into Japan at around the time of the Meiji Restoration in the latter half of the 19th century; however, paradoxically, this importation heightened the consciousness of the Japanese philosophy of the 'unity' (instead of division) of subject and object and led scholars to advocate it as one of the unique features of Japanese philosophy. Throughout the modern period, a considerable number of studies have been conducted to explore the origins of such a philosophy: Some scholars point to naïve animism or indigenous thought based on a sense of unity between humans and nature, whereas others consider the definitive influence of a Buddhist view of the universe, in which humans and nature are one, without any existential distinction. A recent reflection urges a more careful distinction between historical facts and ideological discourses in each thesis (Suzuki 2018). We cannot deny the influence of such discourses, whether historical or ideological, on contemporary Japanese eaters and agri-food professionals.

In addition to this issue, another major challenge concerns how to judge the implication of such a philosophy of unity from the perspective of the totality of food: Should it be regarded as a promising case for realising the affective connection of eaters with contemporary food-related social issues? Or should it be seen as an epistemological obstacle to generating the dialectical dynamism between subjects (eaters) and objects (food and society) to acquire a truly humanistic sense of eating? I refrain from answering this question here and tackle other relevant issues first.

9.1.2 *A Challenge in Defining the Scope of 'Philosophy'*

If a comparative perspective is adopted in relation to French gastronomy, Japanese gastronomy must face the problem regarding the very definition of philosophy. An operational distinction between philosophy and thought would be useful here because there is a persistent view in Japan that philosophy is a systematic body of knowledge imported from the West after the Meiji Restoration. Chōmin Nakae, a modern political theorist and translator of Rousseau's *The Social Contract*, even lamented that 'in Japan there has been no philosophy since antiquity until now' (1901). Certainly, various perspectives can be adopted. In terms of the perspective of world philosophy, any philosophical problem is considered universal to humanity (Nakamura 1998), whereas a more analytical view involves a focus on the Japanese elements within the acculturated Western philosophies (Fujita 2018).

Two implications of this general discussion by Japanese philosophers can be extracted to better analyse Japanese gastronomy. First, speaking of the intellectual realm, the Cartesian exclusion of food from philosophy, which bothered Western thinkers from the 17th century, was also automatically imported into Japanese philosophy; worse, the delayed introduction has left the excluded status of food unaddressed until today.[1] If coupled with the aforementioned argument, it is more precise to state that the Japanese philosophy and thought about food are characterised by the double structure of the dichotomy between subjects and objects in the intellectual sphere and of its unity in the popular sphere. Second, if our goal is to understand the transition of the intellectual status of food, it is more effective to explore not only Japanese philosophy, but also Japanese thought. However, doing so brings with it the complicated challenge of determining the analytical scope of such thoughts on food, such as whether to include ordinary people's thoughts on health (*yōjō*) and even peasant thoughts (*nōmin-shisō*) prior to modernisation, and on what grounds to justify these focuses. The perspectives of peasants and ordinary people are not usually included in Western discourse on food philosophy.

9.1.3 Development of Japanese Food Culture Studies

Indeed, the philosophy of unity has had a non-negligible impact on the development of food culture studies in Japan since the 1970s. Back in the beginning of 20th century, food cultures were already studied by various scholars (e.g., history and folklore studies), among which Kinji Imanishi's (1902–1992) ecological anthropology became the major basis for subsequent food culture studies in Japan. Although Imanishi is internationally recognised as the father of primatology, the epistemological and ontological nature of his ecological anthropology was formed under the strong influences of his passion for alpinism and the Kyoto School philosophy (Nishida 1911), both of which had been historically premised on the Buddhist view of the human-nature unity. Imanishi's (1941) ecological anthropology was later elaborated and applied to food cultures analysis by his successors, notably by Naomichi Ishige. Having started his career as archaeologist, Ishige (1973) combined the ecological perspective with the analysis of material cultures (i.e., cooking utensils and embodied symbols) to formalise the method of food culture studies. As the director of the National Museum of Ethnology, he has facilitated an interdisciplinary dialogue and institutionalised food culture studies in Japan in the form of edited series such as *Shoku no Bunka* (*Food Cultures*: 1998–1999) and *Sekai no Shoku Bunka* (*The World Food Cultures*: 2005–2009). His final publication, *History of Japanese Food Culture* in 2015, was translated into several languages and serves as a referential text for foreign scholars interested in Japanese food cultures.

Keeping this unique academic history in mind, let me summarise its major characteristics from the perspective of the totality of food. Food culture studies in Japan have been institutionalised since early in the 1970s (in parallel with France). However, not having the Annales School or the sociology of eaters, or their equivalents, has resulted in a handicap, namely a non-matured theoretical perspective regarding the totality of food. This is partly due to the fact that Japanese food culture studies originated from ecological anthropology, in which humanity, as Homo sapiens, is conceived as a particular form of animality. This quasi-animistic view is quite different from that of Western naturalism, in which a human-animal existential distinction is advocated (Descola 2013), which, in turn, has also influenced the development of French gastronomy.

As a consequence, Japanese food culture scholars since Ishige's proposal, while pursuing an empiricist and apparently interdisciplinary orientation, have not been so seriously concerned about theoretical and philosophical issues when addressing the total (bio-psycho-social) connection of food and, thus, have been disconnected from relevant international debates, notably regarding the Grand Theories (Mauss, Elias, Lévi-Strauss, Bourdieu, etc.) of food culture studies. Japanese food culture scholars are now compelled to make more self-conscious use of its unique development and the underlying philosophy of unity to articulate universal contributions to these international debates.

9.2 Transition of Culinary Systems and Dominant Social Classes

9.2.1 *Daikyō Cuisine of the Ancient Aristocracy*

French gastronomy was born as an aristocratic practice in court society and, through the dynamism of social distinction, it was transmitted to the lower social classes to finally become a common property of the French. Poulain (2017) stated that 'only Japanese gastronomy in its most sophisticated form, the *kaiseki-ryori* meal style, attains a comparative [with French gastronomy] degree of complexity' (p.183). However, can Japanese gastronomy be considered to have a similar social dynamism? This question is explored in this section by drawing on the general understandings of food historians of the transition of culinary systems in Japan (Harada 2005; Kumakura 2007; Ishige 2015).

In ancient Japan, the establishment of the *ritsuryo* system, a centralised governance structure introduced from China in the 7th century, involved a certain degree of institutionalisation of the Japanese culinary system. In terms of this system, official duties were defined, specifically for procuring foodstuffs (*daizen-shiki*) and cooking (*naizen-shi*), and specific clans were designated to be in charge of these duties, through which a culinary system of rice- and seafood-based dishes evolved by the 8th–9th centuries. *Daikyō* is an aristocratic cuisine that exemplifies the culinary system of this period in its extraordinary form. It was particularly popular at celebrations of the New Year and for political appointments. As shown in Figure 9.1, several noble people gathered around rectangular dining tables but were served individually and ate dishes that they themselves seasoned with salt and vinegar. Their eating utensils were chopsticks and spoons. A wide variety of dishes was served, including mainly dried, raw or conserved products, but each one was simply cooked. The eater's social status was expressed by the number of

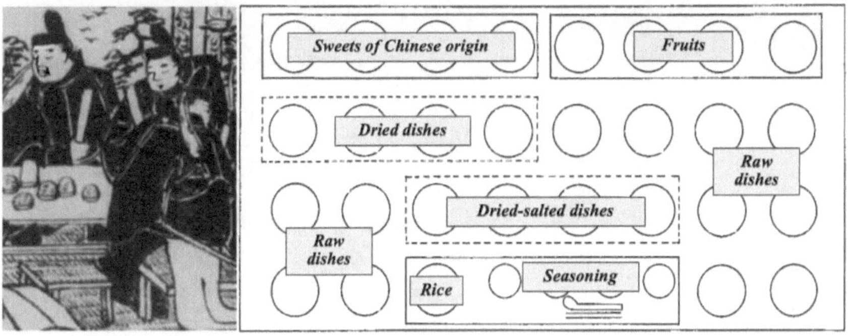

Figure 9.1 Daikyō Cuisine of the Ancient Aristocracy

Source: The left image was cited from *Nenju Gyōji Emaki* (*Scroll Painting of Annual Events*; edited by Y. Tanaka in 1920) in the digital archive of the National Diet Library, Japan. The right image, in which each circle represents a dish, was developed by the author from *Ruiju Zatsuyōshō* (*Book on Court Ceremonies and Furnishings*).

Note: I followed Kumakura's (2007) interpretation of these classical documents in developing this figure.

dishes they received; for example, men of the highest rank were served 28 dishes (Figure 9.1), while those of the lowest rank were given only eight dishes. Rules about table manners were strict, not only in *daikyō* cuisine but also at daily meals, and they functioned as a marker of social distinction from the popular class. This was clearly expressed by a well-known aristocratic female novelist, Seishō Nagon (966–1017/1025), who insulted a carpenter's table manners in Makura no Sōshi (The Pillow Book) as 'vulgar as if they were a cat that comes back home with dirt.'

9.2.2 *Honzen* Cuisine of the *Bushi* in the Middle Ages

Towards the end of the Heian period (794–1185), political governance had gradually been transferred from the aristocratic class to an emerging social class, *bushi* (warriors), who had accumulated social power based on the expansion of private estates and, finally, had established the Kamakura shogunate, a feudal military government (1185–1333). Nevertheless, cultural hegemony was still exercised by the Kyoto-based imperial court. The *bushi* began copying the style of the nobles; however, the content of their meals had to be simplified due to the *bushi*'s insufficient economic means and a resultant asceticism regarding food.

During the following period of the Muromachi shogunate (1338–1573), the social productivity and distribution systems were gradually enhanced and the *bushi*'s gastronomic life became increasingly splendid. It was in this social context that a new culinary style, honzen cuisine, was developed based on the style of *daikyō* cuisine while incorporating the elaborate techniques of *shojin* (Buddhist) cuisine. The most exquisite form of *honzen* cuisine was observed during the *onari*, a special ceremony hosted by a subordinate for his master. *Honzen* cuisine consisted of a drinking part (*kon-bu*) and a meal part (*zen-bu*). The first *kon-bu* started with the

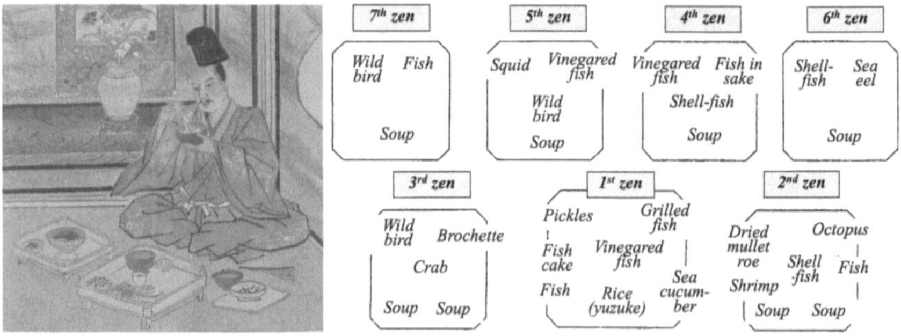

Figure 9.2 *Honzen* Cuisine of the *Bushi* in the Middle Ages

Source: The left image was cited from *Shuhanron Emaki* (*Scroll Painting of Eating and Drinking*) in the digital archive of the National Diet Library, Japan. The right image was developed by the author from *Eiroku Yonen Miyoshitei Onariki* (*Note on Load Miyoshi's Onari* in 1561).

Note: I followed Kumakura's (2007) interpretation of these classical documents in developing this figure.

rite known as *shiki-san-kon*, which involved drinking three cups of sake (Japanese rice wine) with side dishes, and repeating this process three times. The *zen-bu* was composed of several sets of small dining tables called *zen* (for example, three *zens* for the left ceremony and seven *zens* for the right one in Figure 9.2), each of which contained three to five dishes (excluding pickles). The *zen-bu* was followed by the second *kon-bu*, in which several side dishes were served again. Similarly to *daikyō* cuisine, the number of sets of tables and dishes in *honzen* cuisine expressed the social status of each subordinate. The *bushi* enjoyed *honzen* cuisine, not only at *onari* ceremonies, but also at various occasions in simplified forms. It was during this period that some culinary schools also emerged and produced cookbooks, which contributed to normalising this style.

9.2.3 *Shojin* and *Kaiseki* Cuisine

Alongside the *bushi* society, the monk society also developed *shojin* (Buddhist) cuisine. Since the Kamakura period, particularly, Zen monks had introduced from China a sophisticated technique of grinding and stewing vegetables and beans. Through an active interaction between Zen monks and the *bushi*, *shojin* cuisine also influenced both the technical and philosophical aspects of the *honzen* cuisine of the same period. A particular contribution was Dōgen's (1200–1253) *Tenzo Kyokun* (*Instruction to the Monastery Chief Cook*) and *Fushuku Hanpo* (*The Dharma for Taking Food*). Dōgen clearly defined cooking and eating as integral parts of Zen practice and established an aesthetic attitude to food, such as respecting the taste of foodstuffs themselves and a sense of seasons, which were to form the spiritual tenet of *washoku* (as seen in Chapter 6).

Buddhism also contributed to the development of tea drinking. From its introduction to China in the 8th century, the culture of drinking tea was gradually adopted by ordinary people in the Muromachi period, mainly through the tea provided by teahouses in temple towns and the humble houses of Buddhist recluses. Although the upper class advocated more gorgeous types of tea ceremonies that involved *honzen* cuisine and luxurious tea utensils, Sen no Rikyu (1522–1591), who inherited the popular tradition, ultimately aestheticised the humble popular tea ceremony as *wabi-cha*. The *wabi-cha*-style tea ceremony consisted of a simple meal (not more than *ichi-ju-sansai*: rice, soup and three dishes) and tea drinking, the former of which was referred to as *kaiseki*. As J. Rodrigues, a Jesuit priest with a mission in Japan, observed in his *History of the Japanese Church*, *kaiseki* began as a new culinary style in which the taste of foodstuffs, temperature and a sense of season were respected. It is, thus, completely differentiated from *honzen* cuisine.[2]

The early Edo period (the end of the 17th century) saw the transmission of *kaiseki* to the wider population through the replacement of the previously dominant luxurious style of tea ceremony with the *wabi-cha* style, the simplified nature of which indeed had economic merits for emerging merchants. Moreover, by being detached from the strict manners and codes of the *wabi-cha*, *kaiseki* cuisine was made far more accessible to the population in restaurants, the number of which increased immensely in the late Edo period (the beginning of the 19th century). Through the development of tourism and numerous publications of culinary books, *kaiseki* cuisine also started being transmitted to regional cities during the same period.

9.2.4 *Basic Features of the Process of Developing Kaiseki Cuisine*

I have so far explained that *kaiseki* cuisine acquired its complexity from the sophisticated form of the *daikyō* and *honzen* cuisines, the elaborate content of *shojin* cuisine and the tea ceremony backed by the Zen philosophy. When compared with the development of French gastronomy, three basic features of the development of Japanese gastronomy can be identified.

The first relates to the social copying and distinction. Certainly, such social dynamics were partially observed, for example, in the medieval *bushi's* copying of aristocratic *daikyō* cuisine and the Edo merchants' appropriation of *wabi-cha* and *kaiseki* cuisine. However, these examples represent no more than the transmission of dominant culinary systems that automatically corresponded with the transfer of political and cultural power in each period. Thus, these incidences cannot be equated with the dynamism of the court of Versailles and post-revolutionary France. More pertinently, it was not the social distinction across classes but across neighbouring countries (China and later the West) that contributed enormously to the development of Japanese gastronomy, although the role of the former distinction awaits more in-depth analysis in relation to the class structure in Japan.

The second feature is the critical importance of the Zen philosophy. It played almost the same role in changing the Japanese culinary and gustative paradigms as the alchemical philosophy did in France. I discuss this significance in greater detail later in relation to Japanese religions.

The third difference can be observed in the history of transmission. As was discussed at length in Chapters 6–8, French gastronomy, although being partially

Figure 9.3 Ichi-ju-sansai
Source: MAFF (n.a.).

modified in its representation, has been transmitted to contemporary eaters mostly as it used to be. In contrast, *kaiseki* cuisine, after the early-stage expansion in the late Edo period, remained common only in restaurants and is not widely shared by Japanese contemporary eaters in their daily lives. Nevertheless, it is also possible to identify some partial and simplified inheritance of the basic style and spirit of *kaiseki* cuisine to ordinary meals of contemporary Japanese. One such notable case, *ichi-ju-sansai*, a meal composed of rice, soup and three dishes, as depicted in Figure 9.3, is discussed more thoroughly below.

9.2.5 *Ichi-ju-sansai* from the Modern to the Contemporary Period

Until the Edo period, *ichi-ju-sansai* referred to a meal structure as minimal for *honzen* cuisine and maximum for *kaiseki* cuisine in the *wabi-cha*, both of which were limited socially to the upper society and temporally to some special occasions. For ordinary families at the start of the 20th century, the daily meal structure was, at best, *ichi-ju-issai* or *nisai* (rice, soup and one or two dishes), and *ichi-ju-sansai* could be found only at restaurants, such as the cafeterias of department stores (Cwiertka & Yasuhara 2016). Thanks to the development of agricultural production and family cooking facilities during the post-war recovery in the 1950s, the dishes on a family table became richer and more diverse, and *ichi-ju-sansai* for family meals was finally made possible in the 1960s. However, the basic family meal structure was still, at best, *ichi-ju-nisai* during this period, and it was not until the mid-1980s that *ichi-ju-sansai* became normalised for family meals.[3]

Two conditions can explain this sudden change in the status of *ichi-ju-sansai* from an 'ideal and extraordinary' to a 'basic and ordinary' family meal in the mid-1980s. The first condition was the movement of Japanese women out of the traditional housewife role (Ochiai 1994). In the 1980s, the bottom of the so-called M-shaped curve became shorter, indicating the declining tendency of women to quit their jobs when they married and gave birth to their children, and being subsequently re-employed. The accelerated social promotion of women in the 1980s led, on the one hand, to simplifying the dietary norms but, on the other hand, to stigmatising such simplification as a counter-reaction. The latter symptom was caused by (i) the emergence of housewives-turned-cooking specialists who aimed to increase their professional legitimacy and (ii) by the notion that it was the moral responsibility of women to provide family meals, which was promoted as a conservative social psychology in the face of women's increasing social promotion.

The second condition was the increasing impact of nutritionist food discourse. These conversations were started by the US government's McGovern Report in 1978, in which the nutritional balance of a traditional Japanese diet was promoted as ideal for combatting contemporary health issues. Taking this report as an opportunity, numerous dietary guidelines have been issued since the 1980s, including the Japanese Dietary Life in 1983, the Dietary Guideline in 2002 and the Dietary Balance Guide in 2005. Interestingly, the meal structure composed mainly of staple food, a main dish and side dish(es), with no specification of the number of dishes, was normalised in this health-related discourse. Such a meal structure could be easily described as *ichi-ju-sansai* and was closely associated with the symbol of health. The '*ichi-ju-sansai* as healthy' representation was eventually produced and then promoted in the family sphere. Conversely, the reality since the 1980s has been the growth of *ho-shoku*, a destructuring tendency of meals (Adachi 1983; Iwamura 2012) and a widening gap between dietary norms and practices, which causes heavy mental pressure, particularly for working women. Therefore, in articulating the contemporary nature of Japanese gastronomy, it is essential to note the place of nutrition and its social nuances, which are revisited at the end of the discussion of Japanese religions.

9.3 Coexisting Religions and Aesthetics

9.3.1 *Interrelationships of Shinto, Buddhism and Confucianism*

In France, the transformation of the hedonic attitude to food in Catholicism contributed to the consolidation of the spiritual aspect of gastronomy. In Japan, however, a description of religious history itself is too complex, not to mention the place of food in religion. Although I cannot settle conflicting views on Japanese religions here, I follow an orthodox interpretation, as seen in *History of Japanese Religions* (Sueki 2006), to examine the religious condition in relation to Japanese gastronomy.

Before the arrival of Buddhism from China, the indigenous worship of *kami* (divine spirits), which was to later to be called 'Old Shinto', had prevailed widely in ancient Japan. The introduction of Buddhism did not kill belief in these indigenous deities but rather made the Japanese more conscious of them. Accordingly, the syncretism of Shinto with Buddhism was proposed, in which *kami* would coexist with Buddha, but the latter was more politically dominant than the former.

In the Middle Ages, adjacent to or in resistance to the mainstream Buddhism, new Kamakura schools of Buddhism – including Zen schools such as Dōgen – emerged and began to grow in popularity among the Japanese. However, towards the end of the Kamakura period, particularly after the Mongolian invasion (1274, 1281), the political and social presence of Buddhism was gradually weakened. Instead, Shinto became increasingly institutionalised across Japan.

The Edo period is often characterised by the dominance of Confucianism, but, in reality, diverse religions continued to exist and grow in popularity, such as in forms of Buddhist business ethics for merchants and the peasant philosophy, in which the three religions were muddled. Towards the latter half of the Edo period, while these eclectic religious thoughts were rooted in the ordinary people, the *bushi* government made a turn from Shinto-Buddhist to Shinto-Confucianist eclecticism as its authorised religion because the government realised that the increasing nationalism contradicted the foreign nature of Buddhism. In terms of the same logic, the extrinsic nature of Confucianism was also challenged by emerging scholars of classical studies and of ancient Japanese thought, the latter of whom prepared the birth of Reformed Shinto, which served as the theoretical basis of the *sonno-joi* ideology of 'revering the Emperor and expelling the Western barbarians'.

The new Meiji regime began with the Reform Shinto-based policy of separating Shinto from Buddhism, granting the former the status of State Shinto and abolishing the latter. Certainly, the Western principle of freedom of religion was introduced, but the range of religion was strategically narrowed so that State Shinto could be approved as a non-religion and a possible backbone of the political regime. State Shinto and its worship of the Emperor as a divine being began taking root in popular thought, not only through rituals but, most fundamentally, through public education. Under the influence of this ultra-nationalistic State Shinto, no other religions or philosophies (Christianity, Buddhism, Socialism, etc.) could be cultivated. This condition was to terminate at the end of the Second World War.

Beginning in the 1970s, following the setbacks of Marxism, the younger generation started exhibiting an increasing disinterest in socio-political issues and, instead, an interest in new religions. However, the latter tendency was discouraged to some extent because a series of national and global incidents (the 1995 sarin gas attack by members of a Japanese cult movement, the September 11th terrorist attack in 2001, etc.) changed the Japanese representation of religions as dangerous to the social order. Now most Japanese declare that 'they have no religion', but religious tenets are nevertheless deeply rooted in Japanese mores and habits, including food cultures.

9.3.2 *Mixed Nature of Religious History and Discourse*

It would be useful to add two remarks for a better interpretation of Japanese religions. First, despite the seeming diversity and complexity of these religions, factors known as 'indigenous thoughts' (Kato 1975), 'old layers' (Maruyama 1998) or 'old Shinto' in Japanese philosophy studies have played a pivotal role in defining the patterns of accepting or expelling foreign thoughts. These scholars argue that the defining features of such a philosophical basis are *animistic* (founded on the interior unity of man and nature) and *realistic* (non-transcendental, concrete, anti-theoretical and corporeal views of the world), which have modified various foreign thoughts (Buddhism, Confucianism, Marxism, etc.) to make them consistent with Japanese ideas.

Second, historical facts and discourses regarding Japanese religions are muddled. In Japanese philosophy studies, the previous discourse, including the thesis of old layers, has been more or less subject to the socio-political context of each period and has often been confused with historical fact. Nevertheless, the main concern of this chapter is not to verify these historical facts but rather to question the effects of this discourse on Japanese gastronomy. As one of the most representative cases, I discuss the Japanese view of nature and its effect on contemporary Japanese food cultures in the following section.

9.3.3 *Japanese View of Nature*

As pointed out in Chapter 6, respect for nature was defined as an essential spirit of *washoku* as a cultural heritage, but its intellectual origin remained unidentified. This thesis of the Japanese view of nature has faced both criticism and appraisal. It has been dismissed as the reason for blocking the view of nature as an object and, thus, not having developed natural sciences, while it has also been expected as a promising philosophy of the coexistence of man and nature, partly due to the problematisation of the Western mechanistic view of nature in the face of environmental disasters in the 1970s.

Among Japanese philosophers, poetry is considered to reflect the development of this unique view of nature rather well (Kato 1975; Minamoto 1988; Suzuki 2018). The oldest collection of poetry, from the late 8th century, *Manyōshu*, already presented an appreciative view of nature. However, nature was used by the aristocratic poets to merely express romance. The first imperial anthology of Japanese poems, from the beginning of the 10th century, *Kokinshû*, provided a slightly more elaborate view of nature by dedicating a volume to the four seasons, in which the poets no longer extolled human love but preferred to describe its subtle beauties. At the time of the regime change from the Heian to the Kamakura period (at the beginning of the 13th century), the anthology by the impoverished aristocrats, *Shin-Kokinshû*, added a new aesthetic of nature – such as an appreciation of withering of flowers, not the full bloom – by borrowing from the Buddhist awareness of impermanence.

The early influence of Buddhism on the Japanese view of nature can be traced back to the 9th century, when the *hongaku* (innate enlightenment) philosophy of Japanese Esoteric Buddhism was developed (Kato 1975; Minamoto 1988; Sueki 2006). In terms of this *hongaku* perspective, one must realise that all things have the Buddha's nature and, thus, there is an existential unity between man and nature. The Zen philosophy (notably Dōgen) extended this view to be performed in concrete and practical lives (extending food to clothing and shelter), which facilitated its application to the aesthetics of a wide range of arts and techniques. As seen in the previous section, this Zen aesthetics did not exclude the tea ceremony and *kaiseki* cuisine.

In the Edo period, this aesthetic view of nature was completed by Bashō (1644–1694), a poet and lay monk who travelled throughout Japan and witnessed nature with his own eyes, in contrast to previous poets, who were constantly based at the imperial court. Bashō's contribution was significant. In fact, Kato (1975) could not avoid concluding that 'it was through Basho's poems that the Japanese have started to believe they love nature' (p.533). On the other hand, the introduction of Neo-Confucianism infused Japanese intellectuals with the desire to gain a perfect knowledge of natural laws (*kakubutsu-chichi*), which set the stage for the development of practical studies (agriculture, medicine, etc.) in the same period and, ultimately, the acceptance of Western natural sciences in the Meiji period. Since modernisation, two contradictory trends have accelerated the development of a Western scientific view of nature and psychological dependence on the traditional view of unity between man and nature (Minamoto 1988).

Three implications can be extracted for our understanding of Japanese gastronomy. First, again, it was the Buddhist (particularly Zen) philosophy that largely formed the unique Japanese view of nature and its variants – such as the emphasis on freshness, taste and a sense of season in cuisine – and, thus, played almost the same role as alchemy did in the development of French gastronomy. Second, despite its importance, the Japanese view of nature must be interpreted with some reservation because of the limited literary evidence (poetry and arts enjoyed by the aristocracy, *bushi*, affluent merchants, etc.) and the resultant exclusion of more popular and peasant views of nature. For example, to recall one of the features of *washoku*, the 'connection with annual events' is more tightly linked with the latter type of unexplored worldviews than the elite one. Third, the modern sciences have been increasingly influencing or even renewing the Japanese view of nature (Murakami 1977); thus, it is essential to inquire about the degree of such effects. I extend this point by briefly commenting, in the following section, on the place of nutrition in contemporary cuisine in Japan.[4]

9.3.4 *Nutrition as Religion in Contemporary Japan*

In addition to the view of nature, health or nutritional balance was defined as one of the fundamental aspects of *washoku* as a cultural heritage, as seen in Chapter 6. As I have noted throughout this book, this definition was largely due to the nutrition-centred tendency facing contemporary eaters in Japan. Two more examples suffice to illustrate this point. First, the analysis of food discourse in Japan since the 1970s (from the McGovern Report to the Dietary Balance Guide) tells us that they were all focused on meal content and their nutritional advantages, while neglecting the totality of what eating meant to life. Although a partial attempt to include commensality and the pleasure of eating has been observed recently, these were still justified in terms of their nutritional and psychological benefits rather than by socio-cultural values, including socialisation with others (Ueda 2022a). Furthermore, a recent empirical study corroborated that the representation of 'eating well' among contemporary Japanese was heavily focused on health and nutrition, and minimally on the socio-environmental dimension of eating (Ueda 2022a, 2022b).

These examples illustrate how much contemporary Japanese eaters are concerned – somewhat pathologically – with nutrition. The Japanese are acclaimed atheists but, indeed, they choose what and how to eat based on their quasi-religious faith in the nutritional sciences, which I am tempted to call 'nutrition as religion'. To better understand contemporary Japanese gastronomy, it is, thus, essential to examine more closely the historical transitions of Japanese thought on nutrition and health, as well as patterns of how the nutritional sciences have been accepted and modified by indigenous thought.[5]

9.4 Concluding Remarks

The aim of this chapter was to outline a research programme for the sociology of Japanese gastronomy by discussing the philosophical, social and religious conditions of its development. Among these conditioning factors, the long-lasting effect of indigenous and animistic thought and the interrelationships with neighbouring countries (China and the West) played a particularly pivotal role in characterising the historical development of Japanese gastronomy. Although many unsettled issues were presented in this chapter, it is possible to summarise these into two meta-level challenges. First, previous studies of Japanese food cultures were an inquiry merely into *food cultures in Japan*, not *Japanese food cultures*. The consequence of the former neglect of 'Japanese-ness' have facilitated an empiricist focus on accumulating historical evidence regarding foodstuffs, utensils, table manners and culinary styles, without being troubled with the discursive and ideological issues of Japanese philosophy. However, their disconnection from universal questions regarding the view of nature, nutrition and aesthetics, on the one hand, and concerning the theoretical challenges taking place at the bio-psycho-social intersection of eating, on the other hand, are a significant loss to international scholarship. It is tempting to put an end to this *sakoku* (isolationist) policy on Japanese food culture studies.

Second, in advancing such an open and comparative approach to food cultures, one must face a previously underestimated feature of Japanese gastronomy, namely hybridity (Kato 1974). As discussed throughout this chapter, hybridity was both the structured product and the structuring driver of the genesis and transmission of Japanese gastronomy, although it was, unfortunately, masked in the state-run heritisation project of *washoku* due to its ethnocentric characteristics. Researchers who embark on programme aimed at exploring the sociology of Japanese gastronomy need to be conscious of its hybridity and, more ambitiously, should take advantage of the dialectical dynamism of indigenous and foreign elements to produce universal implications for gastronomy studies.

Notes

1 Very recently, some Japanese publications titled 'Food Philosophy' appeared (such as Kawakami 2015; Higaki 2018). However, none addressed such historical complexity regarding the Japanese philosophy of food.

2 For the *wabi-cha*, I refer mainly to a new but more credible narrative (Kozu 2014, 2021) than the popular discourse, with the following two modifications. First, although admitting to two tea ceremony traditions, the interaction between *honzen* cuisine and *kaiseki* cannot be denied. Indeed, *ichi-ju-sansai* (which forms *kaiseki*) is a somewhat simplified version of *honzen* cuisine (or its predecessors, such as *daikyo* cuisine). Second, although acknowledging that the contribution of the Zen spirit was partly exaggerated under the subsequent nationalistic discourses (notably in the 1920–1930s), it is the 'effect' of such discourses that matter in analysing contemporary gastronomy in Japan; thus, the role of the Zen should still not be underestimated.

3 This discourse is based on the content analysis of one of the oldest established magazines, *Eiyo to Ryori* (*Nutrition and Cooking*) in the digital archive of Kagawa Nutrition University (https://www.eiyotoryori.jp/).

4 For further historical review of Japanese view of nature, see Ueda (2022c).

5 One of the possible influences can be seen in the relationship between nutrition education and taste education. In contrast to the complementarity of these two pedagogies in France (Chapter 5), the range of nutrition education in Japan is extended without limit (Adachi & Eto 2005) and the scope of taste education is understood only within a nutrition and health paradigm (as seen in the textbooks such as Ishii et al., 2016). One might be able to identify here the influence of the Japanese animistic and unseparated view of nature and man, or of the biological and social.

References

Adachi, M. (1983) *Naze Hitoride Taberuno [Why Do Children Eat Alone?]*. NHK Publisher.

Adachi, M. & Eto, K. (2005) Expectations for 'Shokuiku' (food and nutrition education and promotion). *Japanese Journal of Nutrition and Dietetics*, 63(4), 201–212.

Cwiertka, K. & Yasuhara, M. (2016) *Himerareta Washokushi [The Hidden History of Washoku]*. Sinsensha.

Descola, P. (2013) *Beyond Nature and Culture*. University of Chicago Press.

Dōgen (1991) Tenzo Kyokun [Instruction to the Monastery Chief Chef]. In S. Nakamura, Ishikawa, R. & Nakamura, T. (eds.) *Tenzo Kyokun & Fushuku Hanpo*, 18–138. Kodansha.

——— (1991) Fushuku Hanpo [The Dharma for Taking Food]. In S. Nakamura, Ishikawa, R. & Nakamura, T. (eds.) *Tenzo Kyokun & Fushuku Hanpo*, 140–234. Kodansha.

Fujita, M. (2018) *Nihon Tetsugakushi [The History of Japanese Philosophy]*. Showado.

Harada, N. (2005) *Washoku to Nihonbunka [Washoku and Japanese History]*. Shogakukan.

Imanishi, K. (1941) *Seibutsu no Sekai [The World of Living Things]*, Kobundo. Translated by P.J. Asquith (2003) *A Japanese View of Nature*. Routledge.

Ishige, N. (1973) *Sekai no Shokuji Bunka [The Food Cultures of the World]*. Domesu Shuppan.

——— (2015) *Nihon no Shokubunka [History of Japanese Food Culture]*. Iwanami Shoten.

Ishii, K., Puisais, J., Sakai, N. & Tajiri, I. (2016) *Kodomo no tameno Mikaku Kyoiku [Taste Education for Children]*. Kodansha.

Iwamura, N. (2012) *Kazoku no Katte desho! [Leave me Alone! Hilarious Family Meals Seen through 246 Photographs]*. Shinchosha Publisher.

Kato, S. (1974) *Zasshu Bunka [Japan as A Hybrid Culture]*. Chikuma Shobo. Translated by K. Yagi and R. Jennison (1986) In *Review of Japanese Culture and Society*, 1(1), 15–24.

——— (1975) *Nihon Bungakushi Josetsu Vol.1 [History of Japanese Literature Vol.1]*. Chikuma Shobo.

Kozu, A. (2005) *Sen no Rikyu no Wabi toha Nanika [What is Sen no Rikyu's Concept of Wabi]*. Kadokawa.

——— (2009) *Chanoyu no Rekishi [The History of Chanoyu]*. Kadokawa.

Kumakura, I. (2007) *Nihonryori no Rekishi [History of Japanese Cuisine]*. Yoshikawa Kobunkan.

Maruyama, M. (1998) Rekishi ishiki no koso [The 'old layer' in history of Japanese philosophy]. In *Chusei to Hangyaku [Loyalty and Treason]*, 353–423. Chikuma Shobo.

Minamoto, R. (1988) Nihonjin no shizenkan [Japanese view of nature]. In S. Omori (ed.) *Sizen to Cosmos [Nature and Cosmos]*, 348–374. Iwanami Shoten.

Ministry of Agriculture, Forestry and Fisheries (n.a.) *Taste of Japan*. https://tasteofjapan.maff.go.jp/jp/topics/detail/134.html (accessed in April 2022).

Murakami, Y. (1977) *Nihon Kindai Kagaku no Ayumi [Japanese History of Modern Sciences]*. Sanshodo.

Nakae, C. (1901) *Ichinnen Yuhan [A Year and a Half]*. Edited by S. Ida (1995) Iwanami Shoten.

Nakamura, H. (1998) *Kodaishiso: Sekai Tsugakushi* [Thoughts in Antiquity: Towards the World Philosophy]. Shunjusha.

Nishida, K. (1911) *Zen no Kenkyu [An Inquiry into the Good]*. Kodokan. Translated by M. Abe & C. Ives. (1990) Yale University Press.

Ochiai, E. (1994) *Kazoku no Sengotaisei no Mikata Koekata [The Japanese Family System in Transition]*. Yuhikaku.

Poulain, J.P. (2017) *The Sociology of Food: Eating and the Place of Food in Society*. Bloomsbury.

Sueki, F. (2006) *Nihon Shukyoshi [History of Japanese Religions]*. Iwanami Shoten.

Suzuki, S. (2018) *Nihonjin no Shizenkan [The Japanese View of Nature]*. Sakuhinsha.

Ueda, H. (2022a) The norms and practices of eating well: In conflict with contemporary food discourses in Japan. *Appetite*, 175, 106086–106097.

——— (2022b) What is eating well? Capability approach and empirical application with the population in Japan. *Appetite*, 170, 105874–105885.

——— (2022c) Japanese view of nature: Discursive tradition, its problems and implications for food studies. *Sustainability*, 14, 8057–8066.

Part III

Theory Construction of Food Education

10 An Integrative Theory for Food Education

Food Ethics and the Capability Approach

Of the two fundamental problems of current food education, the preceding chapters primarily addressed the first type, 'nutritionism', with the help of taste education and gastronomy. It is certain that taste education daringly takes a unique normative approach, called *goût juste*, and thus, more directly deals with the second-type obstacle, the absence of 'ethics', than traditional methods. Nevertheless, a new theory is still required to generalise this particular educational insight to other types of food pedagogies. As discussed in Chapters 1 and 2, various 'theories' have already been proposed, on the one hand, to inform effective food education practices (such as the health belief model and the theory of planned behaviour) or, on the other hand, to establish critical positionings for certain aspects of current food education (notably, Foucauldian discourse analysis and critical theories of neoliberalism). However, none of these theories can both explain the totality of eating lives and articulate the ethical or educational nature of food education. Thus, the aim of this chapter is to construct an integrative theory that satisfies these conditions.[1] To achieve this, I first mobilise 'food ethics' literature and identify the CA as the most promising method.

10.1 Food Ethics and the Capability Approach

10.1.1 Three Major Approaches

Food ethics has been an emerging discipline since the late 1990s. It addresses the ethical aspects of wide-ranging food issues, from meat-eating to biotechnology and business ethics. As in the case of general applied ethics, it is common to begin with the following three major approaches to consolidate food ethics' methodology.

I From the utilitarian perspective, an action is right if it promotes people's pleasure and reduces their pain (Bentham 1789; Mill 1863). A typical example is evident in the argument for veganism: one should abandon the small pleasure of meat-eating to reduce animals' great pain (Singer 1975). A closer inspection of utilitarianism reveals that it can be divided into the following: (1) *act consequentialism* (the goodness of an act is demonstrated by the goodness of its consequent

DOI: 10.4324/9781003341895-14

states of affairs); (2) *welfarism* (the goodness of a state of affairs is demonstrated by the goodness of the utility information regarding that state); (3) *sum ranking* (the goodness of utility information is demonstrated by the total of the utilities in question) (Sen 1985a, p.175). To mention just a few, the utilitarianism poses the following deficiencies: (i) *distributional indifference* (only the sum matters – no matter how unequally distributed); (ii) *neglect of non-utility concerns* (such as rights and freedoms); (iii) *adaptation problems* (one's mental conditioning to difficult circumstances) (Sen 1999a, pp.62–63). At present, there is no utilitarian food ethics research that seriously attempts to answer these criticisms.

II The deontological approach is most fully expressed by Immanuel Kant's (1788) categorical imperative: 'Act only according to that maxim by which you can at the same time will that it should become a universal law'. From this perspective, an action is right only if it is driven by the respect for a rule of conduct (which is duty) that all rational agents should follow. Deontology differs from utilitarianism in two aspects: (1) *non-welfarism* (it is not one's subjective senses but reason that judges the goodness of an action, and following duty is not, thus, always consistent with happiness), (2) *non-consequentialism* (the consequence of an action is merely an accident,[2] and thus, not the basis of moral judgement). In practical settings, however, one duty sometimes contradicts another duty. To resolve this conflict, Kant (1785) introduces a distinction between *perfect* and *imperfect* duties (the latter of which cannot be all the time observed towards all, such as helping others or developing one's talent). Kantian followers find it is useful to specify and compare these multi-layered duties, and this modified approach has been applied in food ethics (Kaplan 2019).

III Virtue ethics used to be the main concern for ancient Greek philosophers, such as Aristotle's *The Nicomachean Ethics*, but had long been overlooked under the supremacy of utilitarianism and deontology. Although visited occasionally by pioneering thinkers (Anscombe 1958; MacIntyre 1981; Nussbaum 1988a, 1988b), it was not until quite recently that virtue ethics was recognised as a legitimate approach. Virtue ethics begins not with the question, 'What is the right thing to do?', but rather with the questions, 'How should one live?' and 'How to make one's life better?' The intellectual gravity is, then, shifted to the characterisation of virtues (such as practical reason, moderation and courage). Although richer and more diverse traditions and methods have been integrated in recent virtue ethics (including Hume, Nietzshe and Confucius), Swanton (2013), in her most comprehensive definition, identifies the three key features of virtue ethics: (1) *Eudaimonistic* (the pursuit of human flourishing), (2) *agent-centred* (neither consequences nor motives), (3) *centred on certain virtue notions*. In addition, Russell (2019) rightly highlights the other critical aspect of virtue ethics:

> What sets virtue ethics apart is that it treats ethics as concerned with one's whole life – and not just those occasions when something with a distinctly 'moral' quality is at stake.
>
> (p.2)

Virtue ethics also marks a critical shift for food ethics, because previous studies (such as Sandler 2014; Thompson 2015) have been more concerned with typically 'moral' food issues (such as meat-eating and food aids) than 'one's whole eating life'. This latter aspect consists of various valuable food-related doings and beings (e.g., what to eat, where and when to eat), and the ethical agents in question are uniquely professionals and particular groups of consumers (such as alternative movements), not ordinary eaters. In this way, food ethics can be a solid basis for inquiring into, 'How should one become a good eater?' (Ueda 2022).

10.1.2 *The Capability Approach as a New Tradition of Virtue Ethics*

Martha Nussbaum was one of the earliest philosophers to realise the connection of virtue ethics with the CA. For Nussbaum (1988a, 1988b), the CA is theorised as being centred on one's capability (virtue) to achieve well-being (eudaimonia), and it integrates the intrinsic importance of agency. To recall Swanton's definition, the CA can, thus, be a form of virtue ethics. Although Nussbaum (2013) interestingly extended the CA to her philosophical characterisation of central human capabilities, it was Amartya Sen who genuinely opened the potential of the CA for empirical application[3] to a variety of social issues, such as poverty, education, health and – which I aim to theorise here – eating lives.

I shall instantly scrutinise his CA's theoretical structure, but here I first present an analytical plan for constructing an integrative theory for food education: (1) The foundational concepts of the CA are revisited and reinterpreted to fit the nature of food education; (2) Sen's own views of 'food' and 'education' are examined to confirm the validity of the first consideration as well as to further clarify the ethical nature of food education; (3) a comprehensive review of CA-based education studies is conducted to explore the appropriate pedagogical nature of food education. Through this three-layered analysis, a set of definitions that stipulate the ontological and epistemological nature of food education are derived to formulate an integrative theory.

To clarify the scope of this theory, food education in this analysis is specified as having the following features: (i) the main target is children. Although I focus mostly on fifth and sixth graders (aged 10–12), who most enjoy food education in Japan, the scope of this theory might not be limited to this range of children; (ii) the main practice field is formal school education.[4] This theory is aimed at understanding neither family education nor informal education (such as TV); (iii) the educational content is concerned with the total range of eating lives, including not only the nutritional, but also the psychological and socio-cultural dimensions. In this sense, food education is distinguished from traditional educational categories (e.g., nutrition education, agricultural education). Fourth, while a maximum effort being made for maintaining the 'positional objectivity' (Sen 1993) of the proposed theory, Japanese examples, to which I am most familiar, are mainly mobilised to derive each definition and demonstrate its implications.

10.2 Foundational Concepts Revisited

The CA is a theory that consists of several analytical concepts, such as well-being and functionings. However, the more CA-applied studies have been carried out, the more diverse interpretations of these very fundamental concepts have been proposed. To clarify the scope of this paper, it is necessary to examine these concepts and redefine them to be aligned with food education.

10.2.1 *Well-Eating and Functionings to Eat Well*

For Sen (1992, p.39), 'well-being' is a person's quality of life. In a food education context, this meaning of well-being is, however, too overarching, since some activities that constitute children's well-being (such as playing with peers, going to piano lessons) might not directly influence their eating lives. Therefore, as far as food education is concerned, it would be appropriate to deal with a more specified aspect of well-being. I call this type of eating life that children value 'well-eating'.

For Sen (pp.39–42), 'functionings' refer to various beings and doings that constitute a person's well-being. Functionings are not merely the beings and doings, but rather the ones the person values (Sen 1985b, pp.19–20). These valuable functionings cover not only elementary levels (such as satisfying nutritional needs) but also more advanced or complex ones (such as joining a community activity) (Sen 1999a, p.5). As in the case of well-eating, I call a specified type of food-related beings and doings 'functionings to eat well'.

Indeed, the strength of the CA is to analytically include the diversity of relationships between commodities and individuals, which was neglected in utilitarian and Rawls' (1971) primary goods approaches (Sen 1980). In other words, the CA seriously takes account of the fact that the conversion of commodities into functionings depends on various personal (e.g., knowledge and education level) and social (e.g., public health facilities) factors (Sen 1985b, p.17). Among these various 'conversions factors', it is important to pay attention to the distinction between choice-free (e.g., food knowledge) and non-choice ones (such as parents' dietary habits) for children. Hereafter, 'commodities' refer mainly to foodstuffs and financial resources (household income or pocket money).[5]

10.2.2 *The Capability to Eat Well*

Having defined each concept, the 'freedom of converting commodities into functionings' is called 'capability' (Sen 1985b, p.17, 1992, pp.110–113, 1999a, p.88). In this sense, capability is a freedom-type concept but, given its inherent conceptual ambiguity, careful consideration should be taken to clarify what Sen actually meant by 'freedom'. He first distinguished the following two meanings (Sen 1985a, pp.208–212, 1992, pp.64–66).

1 *Effective freedom* represents the level of achievement that the person would have chosen. Here, the manner in which one's choice was exercised is not questioned.

2 *Control as freedom* refers to the level of control one is actually exercising. Here, the level of achievement is not questioned.

This distinction is particularly helpful for understanding the confusion facing food education in Japan. This is because there has been much criticism of the Basic Law of Food Education on the grounds that 'food choice is personal and should not be interfered with by the state' (Morita 2004). However, to introduce Sen's distinction, it now becomes clear that these critics perceive ones' food choice only as *control* freedom, while neglecting the possibility that the state can expand their *effective* freedom (food choice) through public policies (notably school-based food education). On the other hand, the state was not aware of such a distinction, and thus could have carelessly violated food choices in the *control* sense. Therefore, in theorising food education, capability should be formulated in a wider sense to integrate both *effective* and *control* freedoms.[6] In a different vein, CA scholars have also clarified at least two senses of Sen's capability (Gasper & Staveren 2003; Qizilbash 2006a).

1 Capability as an *opportunity* to choose a valuable type of life.
2 Capability as an *ability (or skill)* to achieve one's objectives.

To get straight to the point, capability in food education should be formulated from the *ability* perspective, while integrating the initial motivation of the *opportunity* view. The motivation was to take account of social conversion factors influencing children's achievement of given functionings. This is because enhancing their food knowledge and attitudes does not necessarily lead to expanding their freedom to achieve well-eating if the social conversion factors are desperately detrimental to children (such as a lack of access to safe, adequate and healthy foodstuffs in the retail shops).

However, there are two reasons for which viewing capability simply as *opportunity* might not be appropriate in food education. First, even if no active effort is made by children (i.e., personal factors are not improved), their freedom to eat well can be enhanced simply by the improvement of social conversion factors (e.g., deploying more school nutritionists). However, this scenario is not underpinned by the view that children should be active agents (Sen 1992, p.190). Second, the main target of food education is children's own abilities, not their social environments. The latter (environment) will be improved by the former (children).

For these reasons, capability in food education should be conceptualised as *ability*, but at the same time it can entail the motivation of *opportunity*. In other words, this capability is regarded, not as isolated from the food environment, but rather as in a constant interaction with it, which I call 'capability to eat well'. All of these considerations were incorporated into the first definition. It must be noted again that 'capability to eat well' in this sense is wider than what Sen calls 'capability to eat (or to be well nourished)' that has been traditionally used in the context of hunger and famine.

Definition 1. Capability to eat well is the freedom to achieve well-eating (a type of eating life that children value) and is the ability to convert commodities (foodstuff or allowance) into various functionings to eat well. To put it more precisely, this capability is: (1) freedom concerning both achievement and its process; and (2) an ability that is in a constant interaction with the food environment.

This definition has the following advantages for food education research: First, it provides an analytical structure of eating lives, that is, a whole set of relationships of commodity-capability-conversion factors-functioning to eat well (see Figure 10.1 for the visualised structure): To take Sen's example (1985b), individuals with the same amount of food (commodity) might not be able to achieve the same level of well-nourishment (functioning), because their capabilities to eat depend on individually varied personal (e.g., nutritional knowledge, body size, sex) and social (including food education as public policy) conversion factors. This structure is necessary, because previous studies, while aiming at capturing the total range of eating lives, have been unable to propose any evaluative space for the purpose (Adachi 1987) or have managed to point out only some single space (nutrients, foodstuffs or dishes) (Adachi 1984, Toyokawa 2001). In contrast, the first definition orientates us to plurally evaluate each state (functionings, conversion factors, etc.) and its interconnection to better describe the state of one's well-eating.

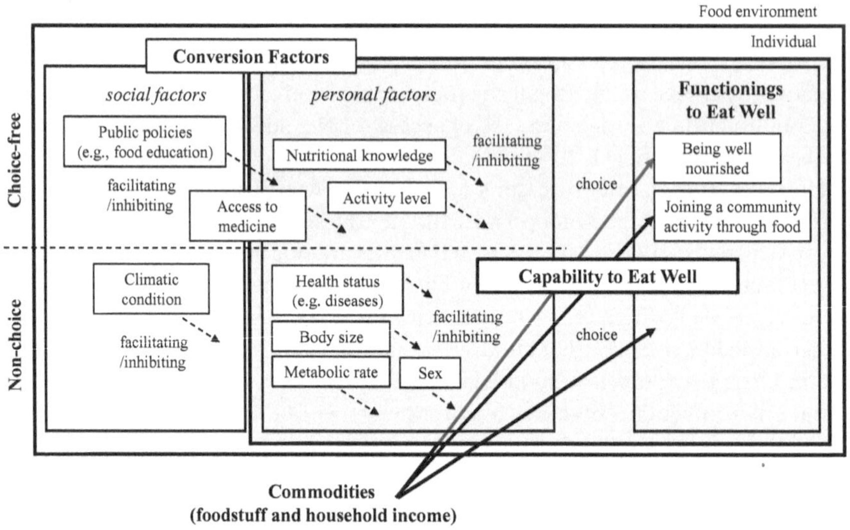

Figure 10.1 The Conceptual Structure of the CA-based Food Education

Source: Reprinted with permission from Ueda (2021, p.8) with a partial revision by the author.

Note: The conversion factors were extracted from Sen (1985b), but these are exemplary and not exhaustive. Arrows from commodities to functionings, which represent the process of conversion, are described in solid lines, whereas arrows from conversion factors to the process of conversion are drawn in dotted lines.

Second, it makes it possible to analytically describe the totality of eating as a set (or list) of functionings to eat well. In previous studies, specific functionings (such as satisfying nutritional needs or knowing where food comes from) were thematised in each educational category (nutrition education, agricultural education, etc.). In contrast, the first definition reminds us that these models are different, simply in terms of what functionings to prioritise, not in its educational nature (as we will see later). In this sense, this first definition provides us with an integrative framework for these existing categories, which would otherwise fall apart or conflict with each other.

Third, it indicates the need for characterising such a list of functionings to eat well for practical purposes. Although such characterisation is not done here, there is a growing body of methodological and empirical studies for such selection and weighting in the CA (such as Robeyns 2003). Nevertheless, the development of the CA-based theoretical framework here is not silent for the content of well-eating. This is because the proposition of this framework itself reminds us that 'eating well' is not just being nourished (as in previous food education) but rather 'eating in a more humanistic way', which can include much wider social and cultural functionings.

10.3 'Food' and 'Education' in Amartya Sen's Philosophy

In the second analysis, I explore how Sen viewed food and education. Although admitting that Sen himself is not necessarily specialised on these themes, at the very first stage of theorising it is important to confirm what elements were already included in Sen's CA and can be further extended by more specialised theories (sociology of food, consumer studies, public health, education theory etc.) in future studies.

10.3.1 Three Meanings of Food: Commodity, Elementary Capability and Choosing

In terms of the food aspect, Sen (1987) describes two scenarios, titled *food for freedom* and *freedom for food*. In the first scenario, having enough foodstuff (as a commodity) and having capability to have enough food are essential to achieve the freedom to live a decent life (such as being free from hunger). In the second scenario, as is the case with Chinese economic reform since 1978, expanding the people's freedom to pursue economic profits leads to expanding food production, and subsequently their capability to be well nourished. To sum up, Sen provides the following two meanings of 'food' that were already included in the first definition.

1 'Food' means a commodity (foodstuff).
2 'Food' means capability to be well nourished, though mainly assuming elementary capabilities (such as nutrition and hygiene) in societies facing imminent famine and starvation.[7]

Sen (1985a, p.212) adds another dimension by saying, 'In matters of what to eat, what to drink [...] it may not be adequate just to get what one would have chosen; one must actually do the choosing oneself'. Although this argument was made as a criticism against the libertarian claim (that advocates control freedom but neglects its consequence), what should be noticed here is that Sen clearly recognises the importance of the 'choice' dimension of eating.

3 'Food' means choosing what and how to eat and also an essential functioning that constitutes one's well-eating. In contrast to the elementary capability above, this choosing also extends to a wider range of functionings to eat well (e.g., choosing a menu in canteens).

As seen in the previous section, this third meaning of food was also already included in the first definition, in which children are recognised as active agents and freedom in the achievement process is catered for. Therefore, any additional definition does not need to be formulated here.

10.3.2 Two Meanings of Education: Intrinsic and Instrumental

In terms of the education aspect, Sen's main perspective was mainly derived from his empirical research on Indian elementary education, in which the following five roles of education were defined (Drèze & Sen 2013).

1 Intrinsic importance: Having education or learning itself is essential for well-being.
2 Instrumental role (personal): Education can expand children's various capabilities.
3 Instrumental role (social): Education can facilitate public discussion and formulation of social needs.
4 Instrumental role (process): The process of schooling is valuable in itself.
5 Empowerment and distributive role: Education enables children of social minorities to resist oppression and to politically organise themselves in order to obtain a fair distribution of resources.

Although these five roles are different in *application*, they are not so in *principle*. Intrinsic importance (1) means that learning (and its joy) in the process of education is a constituent of well-being; that is, it is an elementary functioning. On the other hand, the following three (2–4) are concerned with instrumental roles in expanding some capabilities. However, the social role (3) is the consequence of the personal role (2), since education influences children and these influenced children can then bring about changes in society, not vice versa. Next, the process role (4) is a mixture of two aspects – an instrumental aspect that can facilitate various functionings (e.g., having a wider worldview) and an intrinsic aspect that constitutes the very process of schooling and learning. Finally, the empowerment and distributive role (5) is also a complexity of functionings (such as resisting oppression) and their consequences (such as fairer resource distribution being achieved). In this way, education, in Sen's view, can be formulated more simply in terms of the two roles – intrinsic and instrumental – in principle. These two roles of education can also be correlated with Vaughan's (2007) distinction.

a Intrinsic role of education – education as a functioning (or learning itself)
b Instrumental role of education – education as a (social) conversion factor

From this line of reasoning, the following two provisional definitions can be derived.

Prov. Definition 2a. Food education, if referred to as the process of learning (and its joy), has an intrinsic importance for children's well-eating.
Prov. Definition 2b. Food education, if referred to as a public policy, has an instrumental role as a social conversion factor in expanding children's capability to eat well.

10.3.3 Objective of Food Education

It seems quite natural for most of us to associate food education with the second meaning (2b). In contrast, the first meaning (2a) requires careful consideration. In the first place, by saying 'X is intrinsically important', it means that one's well-being would be unethically damaged without such elementary functioning X (such as having access to basic education and being healthy). If so, it should be questioned whether children's well-being would be unbearably harmed if they did not have food education in schools. At this time, it seems difficult to obtain general consensus regarding the intrinsic importance of food education, since the institutionalisation of school-based food education is quite recent. As was already confirmed elsewhere in this book, most countries had no such food education but nevertheless could not be judged unethical.

Therefore, provisional definition (2a) cannot be immediately integrated into a theory of food education. As a consequence, food education here is regarded uniquely as public policy. Provisional definition (2b) is therefore maintained in the formal second definition. In addition, following Unterhalter's (2003) specification of the differing meanings of education in the CA, it can be considered that this definition refers to the 'objective' of food education.

Definition 2. The objective of food education as public policy (i.e., as a social conversion factor) is to enhance children's capability to eat well.

This definition has the following advantages for food education research: First, it clearly announces the objective of food education. It is contrasted with previous research, in which the objective was formulated in theoretically ambiguous or difficult-to-analyse ways, such as developing 'a rich humanity' (in the Basic Law on Food Education 2005) or 'a holistic understanding of food and nutrition dynamics' (Adachi 2008). Second, it indicates the need to analyse the conditions of public policy to be a promoting social conversion factor (legal framework, deployment of school nutritionists, collaboration between food education NGOs and schools, etc.), which can also be complemented with certain findings from previous food education research (such as Chapter 5).

10.4 Capabilitarian Education and Dewey's Experimental Empiricism

In the third analysis, a comprehensive review of CA-based education studies was conducted. Although many diverse points of discussion were identified, in this section I examine the major three discussion points in order to extract concrete implications for food education.

10.4.1 Quality of Education, Curriculum and Pedagogy

The first of these is related to the quality of education. It is generally agreed that Sen assumed that education could expand children's capability. Therefore, CA scholars pointed out that some types of education could also reduce their capability and it is therefore necessary to scrutinise the quality of education, curriculum or pedagogy (Unterhalter 2003; Walker 2006). Indeed, to complement this argument, it must be noted that even Sen himself paid careful attention to the quality of education by arguing that certain pedagogies that were widely seen in Indian elementary schools (such as rote memorisation and repeating sentences without understanding their meanings) could not expand their capability (Drèze & Sen 2013). In any case, the following definition can be derived from the discussion by CA scholars and Sen.

***Definition* 3**. If an appropriate pedagogy is not employed, food education can reduce children's capability to eat well, although it is not the intention of food education.

This definition has the following advantages for food education research: First, it takes account of both the positive and negative aspects of food education. Previous researchers have been eager to evaluate only positive educational impacts on children (increasing vegetable consumption etc.), while largely neglecting its negative consequences (e.g., limiting their free valuing of food through inculcating single 'correct' eating habits). Second, it demonstrates the need to explore the conditions (or characteristics) for appropriate pedagogies. Indeed, the following fourth and fifth definitions will be a starting point for such exploration. In addition, it also requires a certain level of systematicity (for its educational objective, content and methods) to be named 'pedagogies'. This is suggestive of existing nutrition or agricultural education, which are merely categories, not pedagogies. In this sense, the highly systematic taste education is very close to the philosophy of CA-based food education.

10.4.2 Social Structure, Adaptation and Power Relations

CA scholars, mainly in educational sociology, have pointed out the CA's insufficient attention to social structures and power relations (that construct knowledge in the educational process). By saying insufficient, it does not mean that Sen neglects such social structures (as confirmed by the theoretical importance of social conversion factors) (Robeyns 2005), but rather that some social theories should be added in order to effectively understand these social dimensions. As such, to add-in theories of power relations have attracted growing attention and the so-called 'adaptation' problem in the CA has also been attended to.[8] From the relevant CA studies (Walker 2006; Unterhalter & Walker 2007) and food education ones (Takeda 2008; Hayes-Conroy 2009; Kimura 2011; Leer & Wistoft 2018), three approaches have been proposed for the purpose.

1 To describe the process and mechanism in education whereby children without power (e.g., those of social minority groups) adapt their expectations to their restricted situations and reproduce the existing social inequalities (as advocated by Bourdieu & Passeron [1970]1991).
2 To focus on children's agency and describe the possibility of their resistance and rupture of habitus in education, instead of being obsessed with deterministic reproduction.
3 To employ Foucauldian discourse analysis and critically evaluate the power relations involved in constructing knowledge in schools (i.e., good food, good taste), elucidating diverse negative impacts on children (such as making docile tongues [becoming uncritical about power inequality]).

Since such power inequality is a conversion factor that seriously inhibits children's achievement of given functionings to eat well, it is certain that the language in these theories can reinforce the CA's focus on social dimensions. However, as far as theorising food education is concerned, how to add-in these theories should be carefully considered. This is mainly due to a (seemingly) low affinity of Sen's CA with these postmodernist theories. Particularly among Japanese educational scholars, these postmodernist theories (notably Foucault) have been criticised and even rejected on the basis that postmodernists 'are preoccupied with criticism and cannot construct any educational norm' and 'are leaving education normless and thus paradoxically vulnerable to political and market ideologies' (see Geshi 2016 on this discourse).

From my perspective, these postmodernist theories can be integrated only if they can advance ideas of hypothetical (not absolute) truth and practical reason (or criticism), making them useful for identifying particular power inequalities of given food education. However, given the current theoretical confusion in food education (and education in general) research in Japan, a careless integration would be even dangerous. Rather, education should be first and foremost 'constructive, not critical' (Dewey 1916). Thus, hereafter, I pursue for such a constructive orientation.[9]

10.4.3 Association of Conversion Factors

As a constructive alternative, I focus on John Dewey's education theory. Helpfully, some scholars have already verified the theoretical commonality between Dewey and Sen (Zimmermann 2006; Lessmann 2009), so the theoretical affinity problem is not an issue here. Not just the theoretical affinity, these scholars claim that Deweyan theory can even reinforce some insufficient aspects of Sen's CA, particularly in terms of its social and temporal interaction of agency with the environment. I therefore begin by outlining Deweyan education and extract implications for food education. For Dewey (1938), the theoretical core concept is *experience*. It is not a self-evident concept, and two principles should be referenced to sort out educational types of experience:

1. The principle of continuity, by which the present experience of children is constructed from their past experience and is at the same time constructing their future experience. This principle is a sharp criticism of both traditional education, which overestimates past heritages, and progressive education (as its alternative), which underestimates the past.
2. The principle of interaction, by which objective/external (object, language, facilities, etc.) and subjective/internal conditions (children's interest, ability, etc.) in their experience are constantly interacting with each other. Similarly, this principle refutes both traditional education, which neglects such internal conditions, and progressive education, which underestimates the difficulty in identifying connections of internal and external conditions.

These two principles illuminate Dewey's attitude as a pragmatist philosopher who refuses any artificial dualism (e.g., subject and object) that supports modern philosophy and takes a serious account of the association between things. As a natural consequence, he respects the association between schools and children's daily lives. He then focuses on their daily life experience (such as cooking, gardening and having meals with peers) as ideal educational materials. This is because such daily life experiences more easily induce their interest, which is then transformed into their aims (the result in view). In this process, it is intelligence that can convert their initial interest into their aims. This intelligence can be characterised by certain actions, such as knowledge (from past experience), observation (of the present situation) and judgement (for negotiating knowledge with observation). These intellectual actions are scientific or experimental methods; thus, his theory is called 'experimental empiricism' (Dewey [1916]1975).

To conclude, education in this experimental empiricist sense is defined as 'reorganisation of experience, which adds to the meaning of experience and which increases ability to direct the course of subsequent experience' (p.127). Adding to the meaning of experience here refers to perceiving more and more 'associations' of given actions through which children can better self-control their action to achieve their aim.[10] If the Deweyan concepts (aims, internal/external conditions, association) are translated into the CA's language (functionings, personal/social conversion factors and [partly] social structures), the affinity between both ethical theorists would be clearly observed. The fourth definition was derived from this consideration.

Definition 4. Food education has to include a pedagogical modality that enables children to perceive more and more associations of various conversion factors so that they can better self-control their actions to achieve given functionings to eat well.

This definition has the following advantages for food education research: First, it draws attention to a total range of associations in food education. It is contrasted with previous food education, which, while focusing on physiological associations between children and foodstuffs/nutrients, has not fully analysed their associations with food environments. To put it more practically, most of public policies expect food education in schools to be practiced in a transdisciplinary manner, partly due to the difficulty of setting any food education-specific subject. However, having understood that well-eating cannot be achieved without totally associating knowledge with various subjects (science, mathematics, music, etc.), such transdisciplinary practice has to be more positively justified.[11]

Second, it situates children as subjects who should perceive such associations. Although it is certain that CA-based education scholars have already called for analysing diverse conversion factors, it is generally assumed that these *researchers* are the ones who should perceive such factors, not the *children* themselves. In contrast, from a Deweyan perspective, children must perceive more and more associations for their own development, by which logic conversion factors can be situated *internally* in education itself. Having clarified the right place of conversion factors (and its association) in education, I might be able to say that other social theories (including ones pertaining to power relations) can be effectively added in.

10.4.4 Children's Reason, Subjectivity, Choice and Participation

CA-based education studies were initiated by arguing that the 'CA cannot be applied to children, since their ability to decide is immature' (Saito 2003). Saito's underlying assumption about children's reason and choice has not been critically scrutinised by CA education scholars. Ballet et al. (2011) are among the few who challenged Saito, on the grounds that she did not grasp the importance of children's self-determination.

They went on to make the following argument: In political philosophy (notably Rawls), in which individuals are assumed to be rational and capable of evaluating their own choices, children are given space for only minimal self-determination. It is certain that children cannot enjoy the strong self-determination principle, in which they are considered the 'best judge' of what is good for themselves. In contrast, the weak self-determination principle can be applied, whereby they are regarded as being capable of making choices, but their choice making should be helped to be more reasonable by preparing favourable choice environments (such as eliminating harmful choice options in advance). Having accepted the relevance of the weak principle, more value can be added to children's subjective evaluation, since wanting a snack seems trivial for adults, but might be a big decision or choice for children. In this logic, Ballet et al. situate the importance of children's self-determination, choice, subjectivity and participation.

I fully agree with the importance of these educational measures. Nevertheless, it seems to me that it is also necessary to translate this political philosophy-based logic into that of education and, if necessary, to modify this. First, to be named education, children should be considered capable of making choices and evaluating their choices. Certainly, some of their choices (e.g., eating many snacks) look unreasonable; thus, they need teachers to facilitate the conditions for them to make more reasonable choices (which is so far similar to the weak principle). However, as far as education is concerned, choice should not be viewed as a one-time-only experience, but rather in continuity with their past and future choices (in a Deweyan sense). Even if their present choice ends up being an unreasonable failure, let it be. They can learn from it for the next choice to be more reasonable. If children are regarded as the best judge in this wider sense, the strong principle has to be always applied for education.

Second, it should be noted that in this political-philosophical line of reasoning (including Saito), the immaturity of children's capacity for reasonable evaluation is viewed negatively. However, as far as education is concerned, their immaturity should be interpreted positively as the 'potentiality or capacity to grow' (Dewey [1916]1975). In this sense, Saito's position should be completely reversed into the argument that 'CA can be (or should be) applied to children, since their ability to decide is immature'.

With this modified logic, I now go back to Dewey. Experience in his theory is children's own experience, not the one imposed from outside. Therefore, in the first place, this experience cannot be separated from their subjectivity (including interest) and participation. In addition, for Dewey, education is the process of reorganising experience. In this process, children intellectually convert their interests arising from daily experience (such as cooking) into their aims (functionings to eat well) by using their judgement and performing actions, which are called their 'choices'. In contrast, judgements and actions without such choices would be the ones imposed from outside, or ones in which they unconsciously practice their (bad) habits, both of which should be overcome by education. In this way, the importance of their subjectivity, participation and choice can be more naturally derived from the Deweyan education theory than political philosophy.

Definition 5. Food education has to include a pedagogical modality that respects children's own experience, in which their choice and participation are ensured when converting their subjectivity and interest into given aims (functionings to eat well).

This definition has the following advantages for food education research: First, it indicates the need to reconsider Deweyan experimental empiricism. This also implies a radical challenge of post-war Japanese education. In Japan, Dewey was accepted once as the theoretical foundation to achieve a truly democratic society in the early post-war period. However, it was later the subject of an ideological battle during the Cold War (democratic Deweyan problem-solving versus Soviet systematic approach) and even refuted during the rise of nationalistic revivalism (Sugiura 1998, 2003). In the further centralised educational system (e.g., compulsory educational guidelines and state-authorised textbooks), teachers are facing increasingly more difficulty in exercising their own agency in developing free and creative classes (which are the prerequisite for Deweyan education) (Yamamoto 2014). Therefore, I admit that it is not easy to ground Deweyan food education in current school settings. However, a real expansion of children's capability to eat well cannot be effectively achieved without such experience-based learning, particularly as far as eating habit is concerned.

Second, it highlights the need to be effective for food education by focusing on non-elementary functionings to eat well. The most notable recent example in Japan is a functioning called 'being able to practice traditional food culture'. However, children's values with respect to this might not be the same, depending on how they perceive traditional food culture and how they prioritise this over other functionings. In this case, following the fifth definition, it would be appropriate to keep the *traditional food culture to be transmitted* argument as hypothetical (not fixed) and subject to children's evaluation through choice and participation, through which they could also enhance their ability to engage in open public dialogues (Sen 2009, pp.43–45, pp.324–325).

10.5 Concluding Remarks

The major challenge for existing nutrition-centred food education is the absence of any theoretical framework that can both integrate the totality of eating and deepen the ethical/educational nature of food education. To overcome this issue, this chapter began by reviewing food ethics literature and identified Amartya Sen's CA as most promising for the theoretical development of food education. Through the three-layered analyses, five principles were identified: Food education as public policy has the objective of expanding children's capability to convert commodities into functionings to eat well (Definitions 1 and 2). However, if an appropriate pedagogy is not adopted, food education can reduce their capability (Definition 3). Therefore, the conditions for such an appropriate pedagogy should be explored, and these conditions should, at least, enable children to perceive and act regarding associations of various conversion factors that influence given functionings (Definition 4) and respect their experiences including interest, choices and participation (Definition 5). In this way, the proposed theory can analytically formulate the totality of eating lives as a set of functionings to eat well (Definition 1) and articulate its ethical and educational nature (Definitions 2–5). Thus, the theoretical foundation of food education was established. I move on to discuss this theory's implications in the concluding chapter.

Notes

1 The Sections 10.2–10.4 were reprinted with permission from Ueda (2021). Establishing a theoretical foundation for food education in schools using Sen's capability approach. *Food Ethics*, 6(6), 1–18.

2 According to Kant (1797), one should be always truthful. It would, thus, be one's duty not to tell a lie to a murderer who asked whether his friend, who is being pursued by the murderer, had taken refuge in your house, because doing harm to the friend is merely an accident (for example, even if one told the truth, the murderer might perhaps have been caught by neighbours who came running while he was searching the house for his intended victim).

3 Although Sen is strongly concerned about the connection of economics and ethics (Sen 1982), he constructed the CA's theoretical basis more directly regarding the critical examination of welfare economics and social choice theory (Sen 1999b).

4 At this stage of theoretisation, it would be methodologically dangerous to mix formal education with informal one. This is because, as Dewey (1975) highlighted, the former is 'planned' and requires professionals, pedagogies and institutions, all of which are not the prerequisite for the latter. However, the focus on formal education does not deny the importance of informal pedagogical effects, and the application of Deweyan theory will rather emphasize their connections in the fifth definition, in which children's daily food experience is respected for the purpose of formal food education.

5 Particularly in Japan, 'freedom' has been dominantly interpreted as 'selfishness' due to the Confucianist tradition or, more recently, as 'natural rights' imported from Western philosophy during the early modern period (Wang 2015). It is still quite rare for freedom to be interpreted as in a wide sense as Sen understands the term.

6 After distinguishing two meanings of freedom, Sen prefers to formulate *effective* freedom in a wider sense (i.e., in which *control* can also be considered as a part of the consequence) in order to refute the limited libertarian control view. However, as far as food education is concerned, it would be preferable not to dilute the meaning of *control* (even on the language level) in order to avoid the misunderstanding of freedom by advocates of food education (including the state) or to provide legitimate values for food choices in eating lives. Indeed, Sen (2009, pp.295–296) himself reminds us that the merit of capability is for evaluating the opportunity aspect of freedom, but not necessarily for its process (control) aspect.

7 Sen (1982) most convincingly developed this aspect in his analysis of famine. In this work, his entitlement concept shares its focus on 'relationships between commodities and individuals' with capability, while the latter puts more focus on a wide range of beings and doings (not just having access to commodities as in the former).

8 The adaptation problem is the possibility that people's aspirations and desires are malleable and can 'adapt' in various ways to the straitened circumstances in which they live (Qizilbash 2006b). Due to this innate mental function, if individuals' well-being is being chronically damaged by objective conditions (such as famine), they adapt their desires and expectations to such conditions. Thus, a single focus on the subjective evaluation and its political implication (i.e., no resource allocation to those who are deprived but satisfied with the current conditions) would be problematic.

9 It must be admitted that the discussion here is culturally sensitive and it might reduce the universality of following definitions. Nevertheless, a combination of vigilant stance to postmodernist theories and positive evaluation of Deweyan democratic education is necessary given the historical background of Japan characterised by the ultranationalist educational policy during the World Wars, the post-war integration of Deweyan education under the American occupation (1945–1952), and increasing revivalism of wartime anti-democratic educational policy since the mid-1950s (in fact, until present). For other countries, postmodernist theories could be more easily reconciled with the CA and different definitions concerning appropriate pedagogies could be derived.

10 While Dewey uses 'connection' in his early work, he prefers 'association' in the latter work. Association represents a wide range of connections between humans, knowledge and things. It can refer to political associations, as well as non-political ones (such as friendship), both of which are the prerequisite for democratic societies (Uno 2014). I prefer the association concept which implies the possibility for social reform, to that of connection which is more politically neutral.

11 This relates to a more fundamental problem concerning curricular consistency in current elementary schools. During the early post-war period in Japan, this consistency was supposed to be ensured within children's own experience in the Deweyan 'problem-solving' education. However, along with the rise of economic development since the mid-1950s, it was gradually replaced by 'systematic' education, whereby educational content was upgraded with an uneven emphasis on mathematics and science, while neglecting the importance of curricular consistency (Yamamoto 2014).

References

Adachi, M. (1984) A study on the core-dishes and these combination as a framework of nutrition education on the basis of dish-selecting method. *Japanese Journal of Health and Human Ecology*, 50(2), 70–107.

——— (1987) *Shokuseikatsuron [The Theory of Dietary Life]*. Ishiyaku Shuppan.

——— (2008) Theories of nutrition education and promotion in Japan: enactment of the 'Food Education Basic Law'. *Asia Pacific Journal of Clinical Nutrition*, 17(S1), 180–184.

Anscombe, G. (1958) Modern moral philosophy. *Philosophy*, 33(124): 1–19.

Ballet, J., Biggeri, M. & Comim, F. (2011) Children's agency and the capability approach: A conceptual framework. In M. Biggeri, Ballet, J. & Comim, F. (eds.) *Children and the Capability Approach*, 22–45. Springer.

Bentham, J. (1789) An introduction to the principles of morals and legislation. Translated by Y. Seki (1979) In *Bentham, J.S. Mill.*, 69–210. Chuo Koronsha.

Bourdieu, P. & Passeron J.C. (1970) *La Reproduction : Elements d'une Théorie du Système d'Enseignement*. Editions de Minuit.

Dewey, J. (1916) *Democracy and Education: An Introduction to the Philosophy of Education*. Macmillan. Translated by Y. Matsuo (1975) Democracy and Education. Iwanami Shoten.

——— (1938) *Experience and Education*. Kappa Delta Pi.

Drèze, J. & Sen, A. (2013) *An Uncertain Glory: India and Its Contradictions*. Princeton University Press.

Gasper, D. & Staveren, I. (2003) Development as freedom and as what else? *Feminist Economics*, 9(2–3), 137–161.

Geshi, A. (2016) *Kyoiku shiso no posuto modan [The Postmodern Educational Philosophy: Beyond the Post-war Educational Sciences]*. Keiso Shobo.

Hayes-Conroy, J. (2009) *Visceral Reactions: Alternative Food and Social Difference in Two North American Schools*. Pennsylvania State University, Ph.D. Thesis.

Kant, I. (1785) *Grundlegung zur Metaphysik der Sitten*. Translated by H. Shinoda. (1960) Groundwork of the Metaphysic of Morals. Iwanami Shoten.

——— (1788) *Kritik der Praktischen Vernunft*. Translated by S. Hatano, Miyamoto, K. & Shinoda, H. (1979) Critique of Practical Reason. Iwanami Shoten.

——— (1797) On a supposed right to tell lies from benevolent motives. In T.K. Abbott (ed.) *Kant's Critique of Practical Reason and Other Works on the Theory of Ethics (6th Edition)*, 361–365. Longmans.

Kaplan, D. (2019) *Food Philosophy: An Introduction*. Columbia University Press.

Kimura, A.H. (2011) Food education as food literacy: Privatized and gendered food knowledge in contemporary Japan. *Agriculture and Human Values*, 28(4), 465–482.

Leer, J. & Wistoft, K. (2018) Taste in food education: A critical review essay. *Food and Foodways*, 26(1), 1–21.

Lessmann, O. (2009) Capability and learning to choose. *Studies in Philosophy and Education*, 28(5), 449–460.

MacIntyre, A. (1981) *After Virtue: A Study in Moral Theory*. Notre Dame Press.

Nussbaum, M. (1988a) Nature, function, capability: Aristotle on political distribution. *Oxford Studies in Ancient Philosophy*, Supplementary Volume, 145–184.

——— (1988b) Non-relative virtues: An Aristotelian approach. *Midwest Studies in Philosophy*, 13, 32–53.

——— (2013) *Creating Capabilities: The Human Development Approach*. Harvard University Press.

Mill, J.S. (1863) Utilitarianism. Translated by Y. Seki (1979) In *Bentham, J.S. Mill.*, 459–528. Chuo Koronsha.

Morita, M. (2004) Shokuiku no Haikei to Keii [The background of food education in relation to the Basic Law on Food Education]. *Research and Information*, 457, 1–10.

Qizilbash, M. (2006a) Capability, happiness and adaptation in Sen and J.S. Mill. *Utilitas*, 18(1), 20–32.

——— (2006b) Well-being, adaptation and human limitations. *Royal Institute of Philosophy*, Supplements 59, 83–109.

Rawls, J. (1971) *A Theory of Justice*. Harvard University Press.

Robeyns, I. (2003) Sen's capability approach and gender inequality: Selecting relevant capabilities. *Feminist Economics*, 9(2–3), 61–92.

——— (2005) The capability approach: A theoretical survey. *Journal of Human Development*, 6(1), 93–117.

Russell (2019) Virtue ethics in modern moral philosophy. In D. Russell (ed.) *The Cambridge Companion to Virtue Ethics*, 1–28. Cambridge University Press.

Saito, M. (2003) Amartya Sen's capability approach to education: A critical exploration. *Journal of Philosophy of Education*, 37(1), 17–33.

Sandler, R. (2014) *Food Ethics: The Basics*. Routledge.

Sen, A. (1980) Equality of what? *The Tanner Lecture on Human Values*, 1, 197–220.

——— (1982) *Poverty and Famines: An Essay on Entitlement and Deprivation*. Oxford University Press.

——— (1985a) Well-being, agency and freedom. *Journal of Philosophy*, 82(4), 169–221.

——— (1985b) *Capabilities and Commodities*. North-Holland.

——— (1987) Food and freedom. *Sir John Crawford Memorial Lecture*, 1–25.

——— (1992) *Inequality Reexamined*. Oxford University Press.

——— (1993) Positional objectivity. *Philosophy & Public Affairs*, 22(2), 126–145.

——— (1999a) *Development as Freedom*. Oxford University Press.

——— (1999b) The possibility of social choice. *American Economic Review*, 89(3), 349–378.

——— (2009) *The Idea of Justice*. Harvard University Press.

Singer, P. (1975) *Animal Liberation: A New Ethics for Our Treatment of Animals*. HarperCollins.

Sugiura, H. (1998) *Nihon no Sengo Kyoiku to Dewey [Japanese Post-war Education and Dewey]*. Sekai Shisosya.

——— (2003) *Gendai Dewey Shiso no Sai Hyoka [The Reevaluation of Contemporary Deweyan Philosophy]*. Sekai Shisosya.

Swanton, C. (2013) The definition of virtue ethics. In D. Russell (ed.) *The Cambridge Companion to Virtue Ethics*, 315–338. Cambridge University Press.

Takeda, H. (2008) Delicious food in a beautiful country: Nationhood and nationalism in discourses on food in contemporary Japan. *Studies in Ethnicity and Nationalism*, 8(1), 5–30.

Thompson, P. (2015) *From Field to Fork: Food Ethics for Everyone*. Oxford University Press.

Toyokawa, H. (2001) *Shokuseitaigaku to food system riron [Food ecology and food system theory]*. In H. Toyokawa (ed.) *Shokuseikatu no Henka to Food System [Transformation of Eating Lives and Food System]*, 3–25. Association of Agriculture and Forestry Statistics.

Ueda, H. (2021) Establishing a theoretical foundation for food education in schools using Sen's capability approach. *Food Ethics*, 6(6), 1–18.

——— (2022) Food ethics reexamined: Capability approach and present methodological challenges. *Journal of Ritsumeikan Gastronomic Arts and Sciences*, 7, 251–269.

Uno, S. (2014) *Kaisetsu* [Explanation: Public and its problems]. In J. Dewey (1927). Public and Its Problems. Translated by H. Abe, 307–318. Chikuma Gakugei.

Unterhalter, E. (2003) The capabilities approach and gendered education. *Theory and Research in Education*, 1(1), 7–22.

Unterhalter, E. & Walker. M. (2007) Conclusion: Capabilities, social justice, and education. In M. Walker & Unterhalter, E. (eds.) *Amartya Sen's Capability Approach and Social Justice in Education*, 239–253. Palgrave Macmillan.

Vaughan, R. (2007) Measuring capabilities: An example from girls' schooling. In M. Walker & Unterhalter, E. (eds.) *Amartya Sen's Capability Approach and Social Justice in Education*, 109–130. Palgrave Macmillan.

Walker, M. (2006) *Higher Education Pedagogies: A Capability Approach*. Society for Research into Higher Education and Open University Press.

Wang, X. (2015) Translation enterprises and cultural exchanges between modern China and Japan: The case of liberty. *Bulletin of Institute of Oriental and Occidental Studies*, 48, 174–186.

Yamamoto, M. (2014) *Nihon Kyoikushi [The History of Japanese Education: Thinking Today's Education through History]*. Keio University Press.

Zimmermann, B. (2006) Pragmatism and the capability approach: Challenges in social theory and empirical research. *European Journal of Social Theory*, 9(4), 467–484.

Conclusion

I conclude by answering three inquiries that consistently penetrated this book. First, how the two epistemological obstacles regarding *food* and *education* could be solved? Second, how could food education be constructed as a rival research programme and how can it be advanced empirically? Third, what kinds of 'good eaters' were identified as an educational norm in this research programme?

1 Articulating the Connection of *Food* and *Education*

Since the Enlightenment period until recently, 'food' had not been a legitimate subject in the field of philosophy. On the other hand, 'education' for developing human virtues had been acknowledged by most philosophers; however, almost none seriously deepened this inquiry to concrete theoretical and pedagogical levels. Despite this double exclusion of food and education from intellectual realms, 'food education' emerged recently as one of the most important political and scientific issues. As its historical debt, current food education was to face two fundamental problems, namely its nutritionistic nature and the absence of educational ethics.

1.1 Food as a Total Social Fact

The former obstacle was addressed in multi-faceted manners. Through pedagogical and institutional analyses, taste education was proved to be a promising alternative to traditional nutrition education to develop children's abilities to construct pleasurable relationships with their own senses, others and social environments. This insight was extended by the cultural analysis of gastronomy, identifying the previously neglected link between taste education and this cultural practice. In both analyses, sociology of food played a pivotal role in highlighting the epistemological and ontological natures of taste education and gastronomy. All of these perspectives contributed to enabling us to see 'food' in food education as a total social fact covering the bio-psycho-socio-cultural dimensions of human life.

1.2 Ethics as Capability to Eat Well

Moreover, the inventor of French taste education demonstrated that a philosophical inquiry into this totality of food was to encounter another level of inquiry into

DOI: 10.4324/9781003341895-15

ethical subjects: Taste education is an educational endeavour not to pursue 'good food' but rather to seek for 'good eaters'.

Drawing upon the implication from taste education, the second obstacle of food education was addressed by using ethical theories to food. Through applying Amartya Sen's capability approach and Deweyan educational theory, the ethical nature of food education was ultimately revealed as a tool to expand one's 'capability to eat well', that is, the freedom for achieving a type of eating life (well-eating) that one has reason to value. In this theory, the previously unaddressed concepts in food education, including the well-being, freedom and (in)equality, can be analytically conceptualised.

1.3 *The Ethical and Sociological Approach to Food Education*

This theory construction also presented us further challenges regarding how to identify the diverse functionings as well as the complex 'associations' of conversion factors that influence the achievement of given functionings to eat well. As already exemplified to disclose the total dimensions that constitute French gastronomy (well-eating) in Chapter 7, the sociology of food is a promising candidate, but other add-in social theories are also required to enable a systematic comprehension of functionings and conversion factors (which I shall partly discuss soon later).

The ethical approach is frequently criticised by some sociologists who hold the view that denies any form of individualism. To avoid this confusion, ethical individualism should be distinguished from methodological individualism (Robeyns 2005). It is always each individual that his/her ethics is questioned, but this view does not deny that his/her ethics also depends on social structures; thus, this is where ethics and sociology can complement each other. In this deeper sense, *food* and *education* are inseparably connected within individual eaters.

2 Food Education as a Rival Research Programme

To recall a general scientific history, directing criticism at existing research programmes, such as nutrition and agricultural education, is not sufficient, but it is essential to construct a 'rival programme' with more efficiency in explaining social phenomena in question to move beyond existing programmes. Theoretically, a research programme consists of axioms to define its ontology and epistemology (*hard core*) and complementary hypotheses that protect the hard core from anomalies (*protecting belt*). The hard core of food education, which has been constructed as a rival research programme in this book, ultimately refers to the following definitions, and the protecting belt is played by various add-in theories, including sociology of food.

Definition 1. Capability to eat well as one's freedom to achieve well-eating
Definition 2. The objective of food education as public policy to enhance one's capability to eat well

Definition 3. The negative effect of food education and the demand of appropriate pedagogies

Definition 4. The pedagogies for enabling one to perceive 'associations' of various conversion factors

Definition 5. The pedagogies for respecting one's own 'experience'

These five definitions are not exhaustive and need to be extended with other definitions and add-in theories. Nevertheless, as expressed by Aristotle's optimism in his *The Nicomachean Ethics*, saying 'we have to draw an outline first, and fill it in later. It would seem be open to anyone to take things further and to articulate the good parts of the sketch', I hope such an outline was drawn in the previous chapter. Then, the next issue is how to advance this research programme in the empirical setting. At least three types of tasks are to be addressed in future studies.

2.1 The Content of Well-Eating

First, it would be effective to start this research programme by identifying well-eating (a set of functionings to eat well) in the given context because food education can easily fail if one does not set a goal. The empirical exploration of French well-eating (gastronomy) in Chapter 7 demonstrated one way to proceed with this. Similarly, in Japan, a recent empirical study using the CA revealed 108 kinds of functionings to eat well found in the lay population, covering health (including exercise, sleep, etc.), pleasure (e.g., conviviality, relaxed time at table), regularity (such as having three meals a day), required intake (vegetables, particular nutrients or dishes, etc.), moderation (e.g., not too many calories, or too much sugar and salt), balance (i.e., nutritional) and quality (e.g., fresh, natural, local, home-made) categories (Ueda 2022). Interestingly, although the content of well-eating is, to some extent, different across countries (Lappalainen et al. 1998; Povey et al. 1998; Falk et al. 2001; Guillemin et al. 2016), Ueda's study confirmed that there was also a high level of international universality in content. Figure C1 provides an idea of how the content of well-eating is concretised (although the list is not comprehensive) and its achievement is facilitated by relevant food pedagogies.

Again, the well-eating refers to a set of valuable food-related beings and doings. To apply this logic to existing research programmes discussed in Chapter 1, nutrition education is mainly for achieving eaters' functioning 'to be healthy and nutritionally balanced'; whereas agricultural education is for those 'to create favourable food environments for the access to quality (local, seasonal and natural) foods' through familialisation with agricultural and rural worlds; while home economics education is for those 'to prepare (shop and cook) tasty family meals'; and taste education is for those 'to have gustative and socialised pleasure at table'. Despite the seemingly diverse scopes, all of these, nevertheless, share the same educational foundation for expanding certain aspects of children's capability to eat well. Therefore, they should be complementary rather than conflicting, to achieve the totality of their eating lives. If this rival programme can explain the interrelationship of the existing programmes in this reasonable way, it can be safely said the objective of this book was satisfactorily achieved.

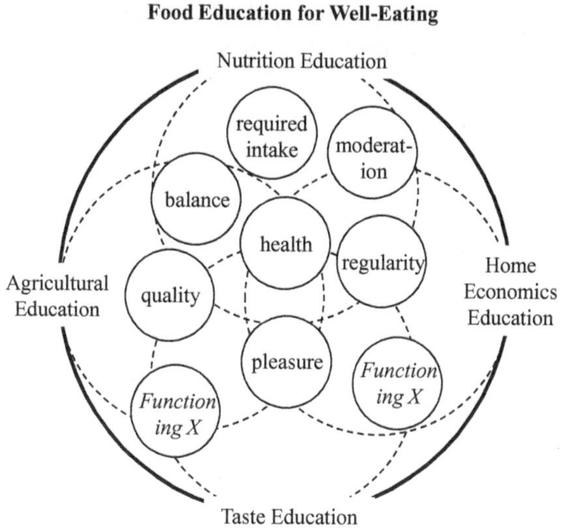

Figure C1 An Example of Well-Eating and Corresponding Food Pedagogies
Source: Developed by the author using the functionings identified in Ueda (2022). The functioning X indicates that the list is incomplete.

2.2 The Association of Conversion Factors and Add-In Theories

Second, it also needs to elucidate the association of relevant conversion factors influencing the functionings to eat well that would be identified in the first task. In addition to the financial, time, motivational and technical constraints, which were partially addressed in Chapter 7 on French gastronomy, there remain more numerous conversion factors. One of deeper-level factors was a cultural norm imposed on contemporary eaters, for which the analysis of French and Japanese gastronomies can serve. In this sense, food education research should be extended to cultural analysis.

There are more influencing factors that have been identified in literature of the nutritional sciences and other disciplines. The challenge in future studies relates to how to integrate these factors into the proposed theoretical framework here, for example, in terms of social psychological tenets (belief, norms, etc.); ecological model and its scales (intra-, inter-personal, community, social, etc.) in nutritional sciences and public health; habits and routines (or habitus) in sociology; and food system and bounded rationality in economics.

2.3 The Pedagogical and Institutional Modality of Food Education

Third, food education as public policy was conceived one of the most promising social conversion factors, and Part I in this book explored the pedagogical and institutional conditions for improving its effectiveness in expanding children's capability. Among these, the development of impact evaluation system is of particular importance in the sense that descriptive nature of given functionings can be translated into their quantifiable forms. As exemplified with taste education in Chapters 4 and 8, the capability for *aliment* was converted to some measurable form, such as the 'knowledge on the taste mechanism' and 'food curiosity', whereas the capability for *convivialité* can also be replaced, partially, with the 'ability for taste expression'. In this sense, the evidence-based food education should be considered as the intellectual exploration for what constitutes people's well-eating and capability, not merely as the (too) technical and meticulous task of effect evaluation (the latter of which is a prevailing view among the social scientists on food education).

In addition to this evidence-based orientation, it is also vital to elaborate food education's pedagogical 'systematicity' that refers to the degree of organisation of education objectives, content and methods based on certain principles. As demonstrated by such a systematic taste education, it can not only reduce the conceptual ambiguity and resultant ideological effects of food education (Chapter 3) but also give an impetus for the sustainable institutionalisation, such as in the forms of the training programmes and of the multi-sectorial participation (Chapter 5). Moreover, it can complement one of the weaknesses of Deweyan educational theory that prioritises the 'experience' of individual children but, in exchange for this, tends to lack concrete methods to valorise such experience to the systematic acquisition of intellectual concepts. This orientation is particularly compelling to Japan, for the food education scholars and practitioners, while eagerly importing foreign (French, Italian, etc.) pedagogies, have yet to develop its own systematic pedagogy. The first step would be to develop 'Japanese' food education with a careful consideration to the uniqueness of its cultural tradition. Understanding food education as each country's 'gastronomic tradition' (not as the product of nutritional evidence) is thus necessary, and this thesis is true of any country beyond Japan and France.

3 'Good Eaters' as Humanist-Democratic Agent

Lastly, I clarify again what kinds of subjects were defined as an educational norm in this food education. The 'good eaters' (gastronomes) are the subjects who carry both the *paideia* and *philanthrōpia* type capabilities. At the first appearance, these subjects seem to be rooted only in the Western traditions (the ancient Greek, Renaissance, Enlightenment, etc.); however, as every country has each own gastronomic tradition, I claim that these are also true of non-Western traditions. Rather, the *paideia* and *philanthrōpia* ethics are more pertinent for some countries like Japan, for these virtues have been less valorised in current food education. Perhaps, more specifications are useful to avoid possible confusion. The *paideia* refers to the knowledge about the total association of man-food-social environments, not merely being knowledgeable about classical texts (as often misunderstood in Japan) (Maruyama 1958). In the context of elementary education, for example, the *paideia* refers to the ability to perceive the transdisciplinary association of food in all relevant subjects.

On the other hand, the *philanthrōpia* refers to the ability for treating others equally (in terms of capability) and constructing pleasurable relationships with them at the table. Given that the philosophy of conviviality was originated from the French democratic tradition of 'Liberty, Equality and Fraternity', the *philanthrōpia* cannot be captured without a solid theory of freedom and equality (that were neglected in nutritional sciences and current food policy). This is also why I was compelled to employ the ethical approach to food education in this book. In sum, eating in a 'more humanistic' way is, thus, a constant effort to know 'more' curiously about food and to achieve 'more' conviviality that is to be made under each individual's structural constraints (time, income, etc.).[1] If humanism is defined in this individually unique sense, these ethics no longer exclude anyone, including the non-Western nations nor the socially disadvantaged in any given region.

So far, good eaters have been conceptualised mainly as 'humanists', however, it also turned out to be 'democratic' agents. This is because good eaters are to develop capabilities to perceive the association of themselves with the social environments (*paideia*) and to cooperate with others through dialogues and discussion (*philanthrōpia*), both of which are also the prerequisite for democratic agents. To conclude this book with Masao Maruyama (1958), leading Japanese political philosopher, who also inspired my analysis of the intellectual condition of Japanese gastronomy, what is desperately missing in nations of unmatured democracy is 'a subject who internally associates radical humanism and radical democracy' (p.179) – if so, one shall also keep striving to become such a humanist-democratic agent in daily eating life.

Note

1 The emphasis on the expression 'more' owes much to Kazuo Watanabe (2005), a contemporary humanist, who as a man of letter strived to promote French humanism to the post-war Japan population without putting its philosophy exclusive to the elite.

References

Aristotle (1971) *The Nichomachean Ethics*. Edited by S. Takada. Iwanami Shoten.

Falk, L.W., Sobal, J., Bisogni, C.A., Connors, M. & Devine, C.M. (2001) Managing healthy eating: Definitions, classifications, and strategies. *Health Education & Behaviour*, 28(4), 425–439.

Guillemin, I., Marrel, A. et al. (2016) How French subjects describe well-being from food and eating habits? Development, item reduction and scoring definition of the well-being related to food questionnaire (Well-BFQ). *Appetite*, 96, 333–346.

Lappalainen, R., Kearney, J. & Gibney, M. (1998) A pan EU survey of consumer attitudes to food, nutrition and health: An overview. *Food Quality and Preferences*, 9(6), 467–478.

Maruyama, M. (1958) *Dearu koto to Suru koto [To Be and To Do]*. Mainichi Newspaper. In *Nihon no Shiso [Japanese Philosophy]* (1998), 154–180. Iwanami Shoten.

Povey, R. Conner, M., Sparks, P., James, R. & Shepherd, R. (1998) Interpretations of healthy and unhealthy eating, and implications for dietary change. *Health Education Research*, 13(2), 171–183.
Robeyns, I. (2005) The capability approach: A theoretical survey. *Journal of Human Development*, 6(1), 93–117.
Ueda, H. (2022) What is eating well? Capability approach and empirical application with the population in Japan. *Appetite*, 170(105874): 1–12.
Watanabe, K. (2005) *France Humanism no Seiritsu [The Birth of French Humanism]*. Iwanami Shoten.

Appendix

Chapter 2

A systematic review was conducted from November to December 2017 and followed the two steps described below (Ueda 2019):

1. *Identification*: This step comprised three tasks. First, the database searching (CiNii Articles for the Japanese articles, Google Scholar for the English ones) was conducted and roughly 8,000 articles including the key term 'food education' were identified. Second, to reduce the number for analysis, the articles of highest relevance (CiNii Articles: all articles published during the past five years [2013–2017], Google Scholar: 200 articles of the highest relevance) were targeted and the journals including these articles were identified as 'food education-related journals'. Third, all articles in the food education-related journals were analysed based on the following eligibility criteria: (i) peer-reviewed research articles and academic monographs whose major focus was (ii) food education in Japan and (iii) 'food education' approached under the emerging conceptual and political framework (meaning traditional nutrition, agricultural and home economics were excluded). After removing duplications, 325 eligible articles were identified.
2. *Categorisation*: The eligible articles were attributed to their corresponding disciplines. The list of disciplines was determined by following the categorisation of the Japan Society for Promotion of Science. The categorisation was processed based on the disciplinary character of the concerned journals (not the respective article) regarding two practical advantages: (i) Some articles included multiple disciplinary approaches or, sometimes, disciplinarily unclear ones, which complicated the categorisation. Compared with articles, 'journals' carry more public and clearer announcement of their disciplinary natures, and thus, they were regarded as the basis of grouping. (ii) Academic journals were one of the key factors for institutionalising the given themes in the concerned disciplines. Although some guest editors from different disciplines were noted, their articles were also counted within the concerned discipline, rather than the original disciplines of the authors. When there was difficulty in categorising, expert opinions on the concerned disciplines were obtained. The list of eligible journals is in Table A1.

Table A1 List of Reviewed Food Education-Related Journals and Eligible Articles

Nutritional and Medical Sciences	
Nutritional sciences Home economics (n = 159)	*Japanese Journal of Food Education* (n = 100), *J. of Dietary Education* (n = 28), *J. of Dietary Life* (n = 14), *J. of Japan Dietetic Association* (n = 7), *J. of Integrated Study of Dietary Habits* (n = 4), *Jap. J. of Home Economics* (n = 3), *Jap. J. of Food Culture* (n = 2), *Jap. J. of Nutrition and Food Service Management* (n = 1)
Social medicine Public health Odontology (n = 58)	*Asia Pacific J. of Clinical Nutrition* (n = 10), *Jap. J. of Public Health* (n = 7), *Jap. J. of Nutrition and Dietetics* (n = 7), *Jap. J. Taste and Smell Research* (n = 6), *Jap. J. of Prevention Systems* (n = 4), *Jap. J. Pediatric Dentistry* (n = 4), *Jap. J. of Health and Human Ecology* (n = 2), *J. of Child Health* (n = 2), *J. Dental Health* (n = 2) *Jap. J. of Hygiene* (n = 2), *Jap. J. of School Health* (n = 2), *International Medical Journal* (n = 2), *Jap. J. of Health Education and Health Promotion* (n = 1), *J. Japanese Association of Rural Medicine* (n = 1) *Jap. J. of Maternal Health* (n = 1), *J. of Nutrition Education and Behavior* (n = 1), *J. of Epidemiology* (n = 1), *Int. J. of Public Health* (n = 1), *Global Health Promotion* (n = 1), *Health Service Insights* (n = 1), *Asia-Pacific J. of Public Health* (n = 1), *Social and Behavioral Sciences* (n = 1), *Maternal and Child Health Journal* (n = 1) *Int. Scholarly Research Notices* (n = 1), *J. Investigative and Clinical Dentistry* (n = 1), *Jap. J. Gnathology* (n = 1)
Social sciences	
Agricultural economics Rural sociology (n = 52)	*Agriculture and Economy* (n = 13), *J. Rural Problems* (n = 7), *J. Food System Research* (n = 4), *J. Rural Life Society of Japan* (n = 4), *Jap. Agricultural Cooperative Research* (n = 2), *Jap. J. Farm Management* (n = 2), *Special Issues of J. Rural Economics* (n = 2), *J. Japanese Cooperative* (n = 1), *Agriculture and Horticulture* (n = 1), *Jap. J. Food, Agricultural and Resource Economics* (n = 1), *Jap. J. Agricultural Education* (n = 1), Monographs (n = 14)
Anthropology Sociology History Business (n = 18)	*Int. J. Business and Management* (n = 2), *Social Responsibility Journal* (n = 1), *Corporate Communication* (n = 1), *J. Social Anthropology and Comparative Sociology* (n = 1), *Asian Politics and Policy* (n = 1), *Int. J. Food Studies* (n = 1), *Social Policy and Society* (n = 1), *Agriculture and Human Values* (n = 1), *Food and Foodways* (n = 1), *Gastronomica* (n = 1) *Studies in Ethnicity and Nationalism* (n = 1), *Journal of Sociology* (n = 1), *Australasian J. Regional Studies* (n = 1), *J. East Asian Studies* (n = 1), *Jap. J. Regional Policy Studies* (n = 1), *Asia Pacific Viewpoint* (n = 1), Monographs (n = 1)

Educational sciences (n = 17)	*Report Studies of Japan Child and Family Research Institute* (n = 3), *J. Childcare Science* (n = 3), *Quarterly Education Law* (n = 2), *J. Family Education Research Institute* (n = 1), *Australian J. Adult Learning* (n = 1), *J. International Student Education* (n = 1), *Jap. J. Educational Technology* (n = 1), *J. Environmental Education* (n = 1), *Jap. J. Home Economics Education* (n = 1), *Jap. J. Agricultural Education* (n = 1), Monographs (n = 2)
Natural sciences	
Agricultural sciences Engineering (n = 21)	*New Food Industry* (n = 8), *J. Nutritional Science and Vitaminology* (n = 3), *J. Food Hygienic Society of Japan* (n = 2), *Foods and Food Ingredients Journal of Japan* (n = 2), *J. Transactions of Information Processing Society of Japan* (n = 2), *Nutrition Today* (n = 1), *J. Japanese Society of Taste Technology* (n = 1), *Food Irradiation* (n = 1), *Nutrients* (n = 1), *J. Industrial Applications* (n = 1)

Chapter 4

The indicators used for measuring concerned educational effects were described below (Tables A2–A4).

Chapter 5

The history of taste education in France was summarised in Table A5.

Chapter 7

Following the method of Lahlou (1995), the protocol used in my analysis to explore the social representation of gastronomy was as follows:

1 The questionnaire started with an open-ended question (Q1) that asked respondents to list the five words associated with 'gastronomy' that sprang to mind. The aim of this question was merely to facilitate the transition to the second main question.
2 Another open-ended question (Q2) was asked to explore the respondents' own definition of gastronomy. The responses are referred to as 'texts', and the whole texts are termed 'corpus'.
3 The corpus was then adjusted into an analysable form through the automatic treatment by IraMuteq as well as the manual treatment by the authors. This treatment included lemmatisation (to reduce the different forms of a word to one single form), stemming (to group different words with similar meaning

Table A2 Adapted Food Neophobia Scale (AFNS)

No.	*Question items*
1	I am very particular about the foods I will eat (reversed scoring)
2	I like foods from different countries
3	I don't trust new foods (reversed scoring)
4	I like to try unusual foods
5	If I have a choice between different flavours for a food (for example: an ice-cream, a sweet…), I like to choose a flavour that I do not know
6	I will try a dish even if I do not know what is in it
7	The foods I know are sufficient for me (reversed scoring)
8	I am willing to eat everything that is offered to me
9	I am afraid to eat things I have never had before (reversed scoring)
10	I will not taste a food when I do not know what it is (reversed scoring)

Note: For questions 2, 4, 5, 6 and 8, the scale ranged from a 'very true for me' response (7 points) to 'true for me' (5 points), 'not true for me' (3 points) and 'not at all for me' (1 point). For questions 1, 3, 7, 9, and 10, the scores were reversed. In this study, AFNS was reduced to eight items, excluding Q1 and Q7, so that the best internal consistency of these items could be obtained (Cronbach's alpha 0.74 at pre-program period).

Table A3 Tentative Food Curiosity Scale

No.	*Question items*
1	Do you like to go shopping with your mom (daddy) to buy food?
2	Do you ask your mom (daddy) to buy food that you like to eat?
3	Do you ask your mom (daddy) to buy food that you like the taste, odour or texture of?
4	Do you want to help your mom (daddy) in the kitchen to prepare the meals at home?
5	When you prepare or when you eat a food that you know, do you love to add new ingredients?
6	Do you like to talk about what you eat even if your mom (daddy) does not ask you questions?
7	Do you like to know what's in a dish (for example, the foods in chocolate, bread or soup)?
8	When you eat at home, do you take the time to look, feel and touch what you are going to eat?

Note: For each question, the score ranged from a 'very often' response (4 points) to 'often' (3 points), 'sometimes' (2 points) and 'never' (1 point).

into one single form [beau and beauté to beau+], elimination of echoes (such as la gastronomie c'est…) and modification of incorrect or abbreviated spelling (such as resto to restaurant).

4 The analysable corpus was statistically treated by descending hierarchical clustering (Reinert 1990; Chartier & Meunier 2011), a programme that systematically classifies the texts according to the analogy and co-occurrence between words.

Table A4 Revised Food Curiosity Scale

No.	*Question items*
1	When I find food that I don't know, I try to ask my parents to buy it.
2	I want to ask my parents to buy foods that have new tastes, flavours or textures.
3	I want to know what ingredients are used in the meal.
4	I find I would rather spend time observing and tasting meals than instantly eating them.
5	It is interesting to learn how food is produced and transported.
6	I find it exciting to know the cultures and histories behind the food that I eat.
7	I try to figure out whether the food is safe by seeing, smelling, touching or tasting it.
8	It is interesting to learn about the relation of food to nature and lives.
9	I like to take care of vegetables in the garden or rice in the paddy.
10	I want to ask my parents to teach me how to cook at home.
11	When I find some fruit or nuts outside, I want to know whether they are edible.
12	I want to make processed foods, such as sausage or miso, from scratch.

Note: For each question, the score ranged from a 'very true to me' response (5 points) to 'true to me' (4 points), 'neither true nor not true (3 points), 'not very true to me' (2 points) and 'not at all true to me' (1 point). These questions should be administered when children are asked to give their responses 'at a particular moment in time.'

Additionally, the confirmative test (of random extraction of 1–5 texts from the corpus) was performed. In nine times of testing, the same classification was produced three times, and similar results of 5–6 semantic classes (in which the current second semantic class was divided or shuffled) were observed for the other six instances. Ultimately, a high level of representativeness of this classification was verified.

The 24 questions in exploring the concrete dimension of gastronomic food model or practice were selected in the following manner:

1 Poulain's (2002b) spatial dimension was used for identifying whether gastronomy can be practiced inside or/and outside the home (Q3) and the reason for this choice (Q4). In fact, this spatial distinction was much contested among the interviewees in the preliminary study, so this distinction was maintained for formulating most of the following questions.
2 Concerning the spatial dimension, two other questions were also integrated (Q10: Where food was procured, Q12: place and type of restaurants). The temporal dimension was used to formulate three questions (Q5–6: timing of meals, Q9: meal duration). The socio-environmental dimension was applied to explore one of the most contested aspects in the interview, in other words, with whom to share gastronomic meals (Q7–8). The three dimensions of Poulain cover the first events or occasions dimension of Warde (2016).
3 Poulain's structural dimension was applied to formulate five questions (Q11, 13: quality of food/restaurants, Q14: recipes, Q15–16: meal components). In particular, the quality dimension had been frequently mentioned but remained conceptually polysemic in the preliminary interview. Therefore,

Table A5 History of the Institutionalisation of Taste Education in France

I Preparatory Stage (1927–1971)	
1927 June	Puisais born into a family of wine merchants in Poitiers
	Obtained a doctoral degree in science from the University of Poitiers
	Started his professional career as a chemist-oenologist
1959	Director at the Departmental Laboratory of Analysis and Research of Indre-et-Loire
1964	Learnt about sensory complexity during the lectures of J Le Magnen, a physiologist of taste
	Established his own experimental laboratory to research the total connection between food and taste
1971	Experimented with taste education for professionals in the wine and culinary industries
	Established the International Academy of Wine as Vice-President
	President of the National Union of Oenologists
II Establishment Stage (1972–1998)	
1972	Started the first taste education experiment for children in Tours, which was later crystalised into ***Classes du Goût***
1974	Established the Conservatoire of Taste
1975	Conducted the *Classes du Goût*, comprising ten sessions, with children in the Centre region
1976	Established the **French Institute of Taste** (the integration of the Conservatoire of Taste), in which interdisciplinary dialogues on taste were advanced
1979	The launch of the sociology of eaters in C. Fischler (ed.) '*La nourriture: Pour une anthropologie bioculturelle*', *Communications*, 31
1981	***'Le vin se met à table'***
	The Socialist Party initiated cultural politics with J Lang (until 1993)
1985	***'Le goût juste des vins et des plats'***
1987	'***Le goût et l'enfant***', the first textbook developed after 10 years of taste education practice
1990	The launch of a national event, *La Journée du Goût* (later called as *La Semaine du Goût*)
	The National Council of Culinary Arts was established as a major funder for taste education
1996	***'Le goûteur et le voluptueux'***
1998	The decomposition of the National Council of Culinary Arts
	(Taste education had been conducted in 3,000 classes with 100,000 children by this time.)
III Incubation Stage (1999–2009)	
1999	'***Le goût chez l'enfant***', the revised version of '*Le goût et l'enfant*' (1987)
	Established the **Institute of Taste** with P MacLeod, a physiologist of taste (the integration of the French Institute of Taste in 2001)
2001	The launch of the **National Nutrition and Health Programme (PNNS)**
	Artistic Cultural Projects initiated by the Ministry of Education, which became one of the main financial sources for taste education
2002	***Classes du Goût*** was upgraded from 10 to 12 sessions
	The Ministry of Education conducted a national survey on the general status of taste education

2005	The National Institute of Agricultural Research (INRA) launched EduSens (until 2009), a scientific project to evaluate the educational impacts of *Classes du Goût*, which constructed evidence for the justification of further political engagement with taste education
2008	The conference organised by the French Institute of Nutrition (IFN), in which the transition from nutrition education to taste education was officially announced
2009	The Ministry of Agriculture set up a working group on taste education to prepare the next development of taste education *(Taste education was increasingly 'scientificied' by the INRA and IFN, politicised by ministries and distinguished from nutritional education.)*
IV Development Stage (2010–present)	
2010 July	The launch of the **National Food Programme (PNA)**, in which taste education was officially positioned as one of the three major axes The Ministry of Agriculture and the Ministry of Education developed a frame of reference, in which the objective of taste education and the qualities of practitioners were defined
Sep	Launched the annual pilot project of *Classes du Goût* in 6 regions with 2,000 children
Nov	Inscription of 'gastronomic meal of the French' as a UNESCO intangible cultural heritage
2011 Jan	***'Et si nous refusions la MacDonaldization du Goût'***
Sep	The Ministry of Agriculture organised a conference titled 'taste education for the young' The popular version of *Classes du Goût*, comprising eight sessions, was developed by the Ministry of Agriculture and the Ministry of Education
Oct	Launched the annual project of *Classes du Goût* in 25 regions and overseas departments The national network of taste education institutions was created by the Ministry of Agriculture (which was upgraded to the National Association of Taste Education for the Young in 2012)
Nov	The Ministry of Agriculture organised a training workshop for taste educators
2013	*(Administrative control was transferred from the central to the regional level.)*
2014	An enactment of the Peillon Law, in which the promotion of **Health Education Guideline (PES)** was stipulated The Law of Future for Agriculture, Food and Forests, in which the implementation of food education in schools was officially stipulated *(Taste education has been promoted up to now in the frameworks of agricultural [PNA], health [PNNS] and educational [PES] policies.)*
2016	Obtained the François Rabelais Prize
2020 Dec	Puisais passed away at the age of 93 due to COVID-19

Source: Ueda (2021a).

a theory of justification (Boltanski & Thévenot 1991) and the convention theory applied in agri-food domains (Allaire & Boyer 1995) were referenced to better orientate the analysis of this quality aspect. In these theories, the quality is not singular and definitive but rather plural and coordinated within different worlds of order (called *cité* in French), each of which has unique criteria for defining its quality: spatial and temporal proximity for the domestic *cité*, originality for the *cité* of inspiration, public good for the civil *cité*, price for the commercial *cité*, efficiency for the industrial *cité*, fame and recognition for *cité* of opinion. The concrete forms and wording of the concerned questions were selected based on previous empirical studies (Andersen 2011; Climent-Lopez et al. 2014; Ponte 2016). Poulain's fifth dimension concerning the logic of choice was not mentioned in the preliminary interview, thus it was not included in this questionnaire. These two dimensions of Poulain cover the second food/menu/dish dimension of Warde.

4 While not directly relevant in the categorisation of Poulain and Warde, the pleasure of eating was frequently mentioned in the interview, thus its meaning and content were further explored (Q17–18).

5 The priority of the abovementioned elements (Q3–18) was also explored to gain a structural vision of gastronomy as an eating model (Q19–20). Further questions were added to explore values attached to such gastronomy (Q21), obstacles for its practice (Q22–25) and actual and ideal frequencies of its practice (Q26).

6 In the third section of the questionnaire, the socio-demographic status of respondents was determined, among which the influences of age, gender, income and education level were analysed in this paper. Using the statistical program R (ver. 3.1.2.), Fisher's exact test for categorical data (Q3, 5–8, 10, 12, 15–18) and Wilcoxon rank-sum test for numerical data (Q9, 11, 13, 19–20, 22, 24, 26) were performed (with Bonferroni correction for comparison among more than two groups).

Index

Note: **Bold** page numbers refer to tables; *italic* page numbers refer to figures and page numbers followed by "n" denote endnotes.